Removal of Pharmaceuticals from Water: Conventional and Alternative Treatments

Removal of Pharmaceuticals from Water: Conventional and Alternative Treatments

Editors

Marta Otero

Carla Escapa

Ricardo N. Coimbra

MDPI • Basel • Beijing • Wuhan • Barcelona • Belgrade • Manchester • Tokyo • Cluj • Tianjin

Editors
Marta Otero
Department of Environment and Planning & CESAM
University of Aveiro
Aveiro
Portugal

Carla Escapa
Applied Chemistry and Physics & IMARENABIO
University of León
León
Spain

Ricardo N. Coimbra
Department of Environment and Planning
University of Aveiro
Aveiro
Portugal

Editorial Office
MDPI
St. Alban-Anlage 66
4052 Basel, Switzerland

This is a reprint of articles from the Special Issue published online in the open access journal *Water* (ISSN 2073-4441) (available at: www.mdpi.com/journal/water/special_issues/pharmaceuticals_removal_from_water).

For citation purposes, cite each article independently as indicated on the article page online and as indicated below:

LastName, A.A.; LastName, B.B.; LastName, C.C. Article Title. *Journal Name* **Year**, *Volume Number*, Page Range.

ISBN 978-3-0365-2457-3 (Hbk)
ISBN 978-3-0365-2456-6 (PDF)

Contents

Preface to "Removal of Pharmaceuticals from Water: Conventional and Alternative Treatments"

Pharmaceuticals represent an especially worrying class of micropollutants because they are biologically active. Thus, their occurrence in aquatic environments may cause undesirable effects in living organisms and, if present in water sources for human consumption, may constitute a public health issue. However, wastewater treatment plants (WWTPs), which have not been designed for the removal of pharmaceuticals and cannot guarantee their full elimination, are considered hotspots for their dissemination in natural waters. In this context, the Special Issue (SI) entitled "Removal of Pharmaceuticals from Water: Conventional and Alternative Treatments" was launched to contribute to the assessment of the contemporary challenges and advances in the removal of pharmaceuticals from wastewater. Papers published in the SI, which have been compiled in this book, approached the topic with either of the following different perspectives: (i) the fate and removal of pharmaceuticals by conventional treatments applied in existing WWTPs; or (ii) advanced and alternative green approaches to remove pharmaceuticals from water. Apart from the Editorial, papers published within this SI include two literature reviews and six experimental studies, all of them presenting unconventional approaches, original views, innovative research and/or novel methodologies. We are grateful to the authors of these publications for their outstanding contributions, and hope that readers may find them useful for their research or in practical applications, and the findings may be shared with colleagues and community stakeholders.

Marta Otero, Carla Escapa, Ricardo N. Coimbra
Editors

 MDPI

Editorial

Removal of Pharmaceuticals from Water: Conventional and Alternative Treatments

Ricardo N. Coimbra [1], Carla Escapa [2] and Marta Otero [1,3,*]

1 Department of Environment and Planning, University of Aveiro, 3810-193 Aveiro, Portugal; ricardo.coimbra@ua.pt
2 Department of Applied Chemistry and Physics, Institute of Environment, Natural Resources and Biodiversity (IMARENABIO), Universidad de León, 24071 León, Spain; carla.escapa@unileon.es
3 Centre for Environmental and Marine Studies (CESAM), University of Aveiro, 3810-193 Aveiro, Portugal
* Correspondence: marta.otero@ua.pt

Abstract: Pharmaceuticals represent an especially worrying class of micropollutants because they are biologically active. Thus, their occurrence in the aquatic environment may cause undesirable effects in living organisms and, if present in water sources for human consumption, may constitute a public health issue. Actually, wastewater treatment plants (WWTP), which were not designed for the removal of pharmaceuticals and cannot guarantee their full elimination, are considered hotspots for their dissemination in natural waters. In this context, the present Special Issue (SI) was launched to contribute to the assessment of the current challenges and advances on the removal of pharmaceuticals from wastewater. The SI consists of seven works with any of two different perspectives: (i) fate and removal of pharmaceuticals by conventional treatments applied in existing WWTP; (ii) advanced and alternative green approaches to remove pharmaceuticals from water. The papers in this SI included five experimental works, two literature reviews, and one case study, all of them presenting unconventional approaches, original views, innovative research and/or novel methodologies. We hope that readers of this SI published by *Water* may find these papers useful for their research or actual activity and may share the findings with their colleagues and community stakeholders.

Keywords: water cycle; emerging contaminants; medicines; antibiotics; pharmaceuticals transportation; water-soluble proteins; microalgae-based wastewater treatment; adsorptive removal; bioremediation; advanced oxidation processes

Citation: Coimbra, R.N.; Escapa, C.; Otero, M. Removal of Pharmaceuticals from Water: Conventional and Alternative Treatments. *Water* **2021**, *13*, 487. https://doi.org/10.3390/w13040487

Received: 30 January 2021
Accepted: 9 February 2021
Published: 13 February 2021

1. Introduction

Freshwater represents around 2.5% of all water on Earth, with less than 1% being accessible. Furthermore, climate change and direct human impacts are dramatically reducing freshwater availability worldwide [1]. Conversely, due to demographic growth, industry development, and the improvement in living conditions, freshwater demand has been continuously increasing. Apart from quantity, water quality is also essential. Indeed, "clean water and sanitation" makes part of the seventeen Sustainable Development Goals by the United Nations. Among the threats to water quality is the occurrence of micropollutants, which include agrochemicals, steroid hormones, personal care products, and pharmaceuticals, which are present at trace levels in the aquatic environment, where they are released by different routes through human activity [2].

In natural waters, pharmaceuticals are micropollutants that mainly come from the after consumption excretion of a non-metabolized fraction and/or metabolites from the original drug. These substances, together with pharmaceuticals wrongly disposed of through the toilet, end up in the municipal wastewater treatment plants (WWTP), which were not designed to remove these sorts of pollutants, but only regulated parameters [3,4].

The nonexistence of discharge limits for pharmaceuticals, which were not traditionally viewed as pollutants, is the main reason for the widespread of these substances. However, their detection in the aquatic environment [2], together with their potential to cause physiological responses in non-target individuals [4,5], raised alarms at the end of the 1990s. Then, pharmaceuticals' detection in drinking water sources made it evident that the effects on human health were not irrelevant [4,6,7].

Concern about the negative effects of pharmaceuticals' presence in the environment has lately led to decisions at a legislative level to determine associated risks. Within the European Union (EU), Directive 2000/60/EC (Water Framework Directive (WFD)) has been the most comprehensive initiative regarding water protection. WFD launched a strategy to define high-risk substances to be prioritized, with 33 priority substances and their corresponding environmental quality standards being ratified by Directive 2008/105/EC. This Directive also set up the establishment of a watch list of 10 substances, in the first instance, which should be monitored across the EU to gather support information for future prioritization exercises. The list was planned to be dynamic and updated every two years so to respond to new information on the potential risks. Then, Directive 2013/39/EU established that the non-steroid anti-inflammatory diclofenac, the synthetic hormone 17-alpha-ethinylestradiol (EE2), and the natural estrogen 17-beta-estradiol (E2) should be included in the first watch list. Accordingly, Decision 2015/495/EU set the definite first watch list, which, besides the referred substances, also contained three macrolide antibiotics, namely azithromycin, clarithromycin, erythromycin, and another natural estrogen, viz. estrone (E1). Decision 2018/840/EU indicated that sufficient high-quality monitoring data were only available on diclofenac, which was removed from the list; the rest of the pharmaceuticals remained on the second list, which also added the antibiotics amoxicillin and ciprofloxacin. Recently, Decision 2020/1161/EU established that, since four years is the maximum that any substance may be on the watch list, EE2, E2, E1, and macrolide antibiotics should be removed while amoxicillin and ciprofloxacin should be maintained in the third watch list. This Decision, in agreement with the EU Strategic Approach to Pharmaceuticals in the Environment and with the European One Health Action Plan against Antimicrobial Resistance (AMR), also set the inclusion of the sulfonamide antibiotic sulfamethoxazole, the diaminopyrimidine antibiotic trimethoprim, the antidepressant venlafaxine together with its metabolite O-desmethylvenlafaxine, and a group of ten azole pharmaceuticals.

The abovementioned initiatives point to the importance of research on suitable approaches to reduce the entrance of pharmaceuticals into the environment. In general, upstream (before release) and downstream (after release) strategies may be adopted for pollution control, with the first being preferred and more effective. However, in the case of medicines, although some improvements may be made before release, namely concerning manufacture, distribution, prescription, consumption or management, and albeit education campaigns are very important [5], restrictions cannot be applied in the same way as for other pollutants since pharmaceuticals are essential to satisfy the population's health care needs. Not to mention that owing to the global population growth and aging, their consumption has an increasing trend [3,8]. Therefore, apart from the upstream strategies, feasible approaches are necessary for the efficient removal of pharmaceuticals from water, which constitutes an actual and great challenge for researchers and engineers working on wastewater treatment.

In the described context, this Special Issue (SI) aimed to provide a platform for scientists to bring forth abatement strategies and treatments, either conventional or alternative, for the removal of pharmaceuticals from water and to discuss treatments' efficiency and the fate of this sort of pollutants.

2. Overview of the Special Issue

Seven high-quality works were published within the SI on "Removal of Pharmaceuticals from Water: Conventional and Alternative Treatments", which consisted of two review papers [9,10] and five research manuscripts covering a wide range of topics related to pharmaceuticals' pollution, fate, abatement strategies, removal treatments, and/or efficiency assessment [11–15].

The two review articles in the collection dealt with two different types of treatment applied for the removal of pharmaceuticals in general [9], or antibiotics in particular [10]. Silva et al. [9] made a wide review of the literature about the use of biological matrices, namely algae and fungi, with a special focus on bioremediation and biosorption. The authors highlighted the advantages of these treatments, such as the low capital investment and the simple and relatively cheap operation. The use of fungus and microalgae for pharmaceuticals' bioremediation, specifically named mycoremediation and phycoremediation, was thoroughly reviewed, covering aspects such as treatment systems, removal mechanisms, factors influencing degradation capability, and future challenges. As for biosorption, advantages, such as avoiding nutrient supply and the generation of transformation products, were pointed out. Fungus and algal cells for the biosorption of pharmaceuticals, treatment systems, removal mechanisms, influencing factors, and biosorption potential were reviewed. Final remarks pointed out that most of published literature deals with laboratory-scale works and synthetic aqueous media so real applications need to be studied; research into the mechanisms and the dependence on physicochemical and biological factors is still necessary; and, finally, genetic engineering should be considered to select the most efficient strains or to modify fungi and algae to be more efficient [9]. For their part, Cuerda-Correa et al. [10] provided a comprehensive review on advanced oxidation processes (AOP) applied for the removal of antibiotics, which were selected as targets due to their recalcitrant properties and actual concern about antimicrobial resistance. AOP were presented as new, sustainable, and clean water purification technologies with large versatility and a broad spectrum of applicability. Catalytic and non-catalytic processes are included within AOP, which stand on the high oxidizing capacity of the hydroxyl radical, differing in how this radical is generated. In this sense, published studies on photolysis, ozone (O_3) based, hydrogen peroxide (H_2O_2) based, heterogeneous photocatalysis, sonochemical, and electrooxidative AOP were reviewed. For each group, different treatments for antibiotics' removal were presented, with the corresponding mechanisms stated, their efficiencies reported, and specific remarks on their performance discussed. As the previous review [9], this also indicated that most published literature was on bench- or pilot-scale studies, with the implementation of AOP at full-scale, still being quite limited [10]. The main goal regarding AOP is to lower overall cost per unit mass of pollutant that is removed, for which achievement, Cuerda-Correa et al. [10] presented three main challenges, namely avoiding unnecessary expenses, reducing energy consumption, and minimizing resulting wastes, and gave suggestions to achieve them.

Among the research articles included in the SI, two of them [11,12] were related to the fate of pharmaceuticals under conventional WWTP. Al-Qaim et al. [11] analyzed nine pharmaceuticals in Malaysian surface water, sewage treatment plant (STP) influent, STP effluent, and hospital effluent. For this purpose, a single solid-phase extraction followed by an accurate and selective liquid chromatography-time of flight/mass spectrometry (LC-ToF/MS) method was developed. The studied pharmaceuticals were atenolol and metoprolol (β-blockers), acetaminophen (analgesic), caffeine and theophylline (stimulants), sulfamethoxazole (antibiotic), prednisolone (steroidal anti-inflammatory), ketoprofen (non-steroidal anti-inflammatory), and glibenclamide (antidiabetic), which were shown to have different doses and consumption patterns. Quantification limits of the developed methodology in STP influent, STP effluent, surface water, and drinking water samples respectively averaged 29, 16, 7, and 2 ng L^{-1}. The most frequently detected pharmaceuticals were nonprescription, namely acetaminophen (75%), theophylline (100%), and caffeine (83.3%), in which respective mean concentrations were 74, 38, and 540 ng L^{-1}. In addition, atenolol, metoprolol, acetaminophen, caffeine, theophylline, and sulfamethoxazole were detected in surface water, STP influent, and STP effluent, with lower mean concentrations in STP effluents than in STP influents, indicating that they were partly removed in the oxidation ditch of the STP. Regarding hospital effluents, pharmaceuticals' mean concentrations were all higher than in STP effluents, which authors related to the relatively low removal efficiency of the rotating biological contractor in the hospital WWTP. Al-Qaim et al. [11] highlighted that the highest determined concentrations were those of caffeine and acetaminophen (8700 and 4919 ng L^{-1}, respectively) in STP influent, which was linked to their high consumption patterns. As for Hofman-Caris et al. [12], they provided an extensive case study on the origin, fate, and control of pharmaceuticals in the river Meuse (The Netherlands), its tributaries, and a drinking water treatment plant (DWTP) downstream. In this work, and for the very first time, after determining the concentration of pharmaceuticals and metabolites in the tributaries, their apportionments to the DWTP intake were estimated and then verified. This relevant and integrative study comprised four steps: (i) compilation of pharmaceuticals and metabolites concentration in effluents from several WWTP; (ii) assessment of loads in the river Meuse and tributaries, with apportionment of WWTP contributions to the DWTP intake; (iii) evaluation of abatement options, including drinking and wastewater treatments; and (iv) presentation of short and long term solutions. Large divergences were found between different WWTP regarding pharmaceuticals' concentrations, with some of them being extremely difficult to remove (for example, diatrizoic acid, metoprolol, or diclofenac). WWTP were proved to contribute to pharmaceuticals loads in the river Meuse significantly. Contributions of the Meuse and tributaries to the concentration of pharmaceuticals and metabolites in the intake of the DWTP were estimated, evidencing that the river Meuse had the largest input. Authors proposed abatement options at WWTP along the tributaries and in the DWTP, indicating their effectiveness, costs, advantages, and disadvantages. Adding refinement treatments able to remove pharmaceuticals at every WWTP in the catchment was highlighted as the best option, also granting good chemical and ecological status in the river and making additional treatments in the DWTP dispensable. In this sense, incorporating an AOP, namely UV/H_2O_2, after removing organic matter by ion exchange was presented as an efficient alternative at a fair spending (increasing costs by about 0.23 € m^{-3} of treated water).

The other three research articles of the SI [13–15] are experimental works on alternative treatments for the removal of pharmaceuticals and with a special focus on sustainability. Coimbra et al. [13] studied the biosorption of diclofenac, a nonsteroidal anti-inflammatory drug included in the first watch list (Decision 2015/495/EU), onto residual microalgae biomass of two different genera, viz. *Synechocystis* sp. and *Scenedesmus* sp. As highlighted by the authors, the implementation of microalgae systems for CO_2 fixation is limited by high costs, which may be reduced using wastewater as culture media, namely, as a source of water and nutrients. To further increase these systems' sustainability and in line with the circular economy paradigm, the residual (dead) microalgae biomass must be given a use. Its utilization as a pharmaceuticals' biosorbent was proposed, and both kinetic and equilibrium experiments were carried out under batch operation. Obtained results respectively fitted the pseudo-second kinetic order and the Langmuir isotherm. *Synechocystis* sp. and *Scenedesmus* sp. biomasses showed similar kinetic performance, with *Scenedesmus* sp. biomass attaining higher diclofenac sorption capacity at equilibrium (28 mg g^{-1}) than *Synechocystis* sp. biomass (20 mg g^{-1}). These values were shown to be lower than by commercial activated carbon but comparable to published results for waste-based activated carbons even when microalgae biomass was not chemically nor thermally modified or treated before use [13]. Thus, microalgae biomass application as a biosorbent was highlighted as a sustainable alternative, also favoring the zero-waste cultivation of microalgae. Kebede et al. [14] carried out a novel study on the biosorption of antibiotics, viz. sulfanilamide, marbofloxacin, ciprofloxacin, danofloxacin, oxytetracycline, sulfadimethoxine, sulphacetamide, sulfamonomethoxine, sulfamethoxazole, tylosin, and sulfamerazine, using water-soluble proteins from the seeds of *Moringa stenopetala*. Moreover, the surface functional groups of water-soluble protein powder before and after adsorptive use were determined by Fourier transform infrared (FTIR). Under optimized conditions, the simultaneous removal of selected antibiotics from synthetic and real wastewater was investigated. Maximum removals in the range of 85%–96% were determined, which decreased to 70%–82% in real wastewater samples. The authors pointed out that the proposed treatment was simple, cost-effective, environmentally friendly, and easily applicable, with *Moringa stenopetala* cultivation contributing towards deforestation reduction. Finally, Escapa et al. [15] assessed the efficiency of different microalgae strains, namely *Chlorella sorokiniana* (CS), *Chlorella vulgaris* (CV), and *Scenedesmus obliquus* (SO), in the treatment of water contaminated with acetaminophen, which is a widely used non-prescription analgesic and antipyretic. As remarked by authors, microalgae may be used for green eco-friendly water treatment, with assets such as photoautotrophic growth, few operational requirements, CO_2 fixation, and generation of both high-value sub-products and profitable biomass. In this work, water was treated under batch conditions in bubbling column photobioreactors run at a semi-pilot scale. Accounting for the possible generation of transformation products from acetaminophen biodegradation, the treatment efficiency was not only determined in terms of removal but also toxic effects on zebrafish (*Danio rerio*) embryo, namely at the gastrula, pharyngula, larval and juvenile stages. At the end of the batch, acetaminophen concentration decreased by an average of 67%, 39%, and 17% under the cultivation of CS, SO, and CV, respectively. In the same way, the incidence of toxic effects on zebrafish embryos was CS < SO < CV, which confirmed CS as the most efficient. Moreover, toxic effects determined for microalgae treated effluents were equal to those of synthetic solutions with equivalent acetaminophen concentrations, which allowed the conclusion that acetaminophen biodegradation by CS, SO and CV did not result in toxic transformation products for zebrafish embryo.

Globally, this SI on "Removal of Pharmaceuticals from Water: Conventional and Alternative Treatments" pointed out the interest of the scientific community about finding efficient treatments and strategies for the removal of pharmaceuticals from water. Apart from such a challenge, this SI has also reflected researchers' concern about the treatments' sustainability from both the environmental and economic points of view and the importance of implementing integrated approaches. Furthermore, it is underlined that AOP and bioremediation received special attention in this SI, having been featured as promissory treatments in the articles here gathered.

Author Contributions: M.O., C.E. and R.N.C., as Guest Editors of the Special Issue entitled "Removal of Pharmaceuticals from Water: Conventional and Alternative Treatments", contributed to the preparation of this Editorial. Conceptualization, M.O., C.E. and R.N.C.; writing—original draft preparation, R.N.C., C.E. and M.O.; writing—review and editing, M.O.; Supervision, M.O. All authors have read and agreed to the published version of the manuscript.

Funding: Thanks are due to the Portuguese Fundação para a Ciência e a Tecnologia/Ministério da Ciência, Tecnologia e Ensino Superior (FCT/MCTES) for the financial support to the Associated Laboratory CESAM (UIDP/50017/2020+UIDB/50017/2020) through national funds. Funding through the FCT Investigator Program (IF/00314/2015) is also acknowledged.

Institutional Review Board Statement: Not applicable.

Informed Consent Statement: Not applicable.

Data Availability Statement: Data sharing is not applicable to this article.

Acknowledgments: The contributions to this Special Issue (SI) are acknowledged with thanks. We would like to express our appreciation to the authors of these contributions for their commitment and superb work. Thanks are also due to the anonymous reviewers who gently revised the submissions to this SI for their time and selfless dedication. Finally, our most sincere gratitude to the editorial managers, especially to Fionna Fu, who so kindly assisted the assembling of this SI. As guest editors, we are very happy with the final result.

Conflicts of Interest: The authors declare no conflict of interest.

References

1. Rodell, M.; Famiglietti, J.S.; Wiese, D.N.; Reager, J.T.; Beaudoing, H.K.; Landerer, F.W.; Lo, M.H. Emerging trends in global freshwater availability. *Nature* **2018**, *557*, 651–659. [CrossRef] [PubMed]
2. Warner, W.; Licha, T.; Nödler, K. Qualitative and quantitative use of micropollutants as source and process indicators. A review. *Sci. Total Environ.* **2019**, *686*, 75–89. [CrossRef] [PubMed]
3. Pereira, A.; Silva, L.; Laranjeiro, C.; Lino, C.; Pena, A. Selected pharmaceuticals in different aquatic compartments: Part I—Source, fate and occurrence. *Molecules* **2020**, *25*, 1026. [CrossRef] [PubMed]
4. Santos, L.H.M.L.M.; Araújo, A.N.; Fachini, A.; Pena, A.; Delerue-Matos, C.; Montenegro, M.C.B.S.M. Ecotoxicological aspects related to the presence of pharmaceuticals in the aquatic environment. *J. Hazard. Mater.* **2010**, *175*, 45–95. [CrossRef] [PubMed]
5. Courtier, A.; Cadiere, A.; Roig, B. Human pharmaceuticals: Why and how to reduce their presence in the environment. *Curr. Opin. Green Sustain. Chem.* **2019**, *15*, 77–82. [CrossRef]
6. Yang, Y.; Ok, Y.S.; Kim, K.H.; Kwon, E.E.; Tsang, Y.F. Occurrences and removal of pharmaceuticals and personal care products (PPCPs) in drinking water and water/sewage treatment plants: A review. *Sci. Total Environ.* **2017**, *596–597*, 303–320. [CrossRef] [PubMed]
7. Luo, Y.; Guo, W.; Ngo, H.H.; Nghiem, L.D.; Hai, F.I.; Zhang, J.; Liang, S.; Wang, X.C. A review on the occurrence of micropollutants in the aquatic environment and their fate and removal during wastewater treatment. *Sci. Total Environ.* **2014**, *473–474*, 619–641. [CrossRef] [PubMed]
8. Quesada, H.B.; Baptista, A.T.A.; Cusioli, L.F.; Seibert, D.; de Oliveira Bezerra, C.; Bergamasco, R. Surface water pollution by pharmaceuticals and an alternative of removal by low-cost adsorbents: A review. *Chemosphere* **2019**, *222*, 766–780. [CrossRef] [PubMed]
9. Silva, A.; Delerue-Matos, C.; Figueiredo, S.A.; Freitas, O.M. The use of algae and fungi for removal of pharmaceuticals by bioremediation and biosorption processes: A review. *Water* **2019**, *11*, 1555. [CrossRef]
10. Cuerda-correa, E.M.; Alexandre-franco, M.F.; Fern, C. Antibiotics from Water. An Overview. *Water* **2020**, *12*, 1–50.
11. Al-Qaim, F.F.; Mussa, Z.H.; Yuzir, A.; Abu Tahrim, N.; Hashim, N.; Azman, S. Transportation of different therapeutic classes of pharmaceuticals to the surface water, sewage treatment plant, and hospital samples, Malaysia. *Water* **2018**, *10*, 916. [CrossRef]
12. Hofman-Caris, R.; ter Laak, T.; Huiting, H.; Tolkamp, H.; de Man, A.; van Diepenbeek, P.; Hofman, J. Origin, fate and control of pharmaceuticals in the urban water cycle: A case study. *Water* **2019**, *11*, 1034. [CrossRef]
13. Coimbra, R.N.; Escapa, C.; Vázquez, N.C.; Noriega-Hevia, G.; Otero, M. Utilization of non-living microalgae biomass from two different strains for the adsorptive removal of diclofenac from water. *Water* **2018**, *10*, 1401. [CrossRef]
14. Kebede, T.G.; Dube, S.; Nindi, M.M. Removal of multi-class antibiotic drugs from wastewater using water-soluble protein of moringa stenopetala seeds. *Water* **2019**, *11*, 595. [CrossRef]
15. Escapa, C.; Coimbra, R.N.; Neuparth, T.; Torres, T.; Santos, M.M.; Otero, M. Acetaminophen removal from water by microalgae and effluent toxicity assessment by the zebrafish embryo bioassay. *Water* **2019**, *11*, 1929. [CrossRef]

Review

The Use of Algae and Fungi for Removal of Pharmaceuticals by Bioremediation and Biosorption Processes: A Review

Andreia Silva, Cristina Delerue-Matos, Sónia A. Figueiredo and Olga M. Freitas *

REQUIMTE/LAQV, Instituto Superior de Engenharia do Porto, Rua Dr. António Bernardino de Almeida 431, 4200-072 Porto, Portugal

* Correspondence: omf@isep.ipp.pt

Received: 3 July 2019; Accepted: 25 July 2019; Published: 27 July 2019

Abstract: The occurrence and fate of pharmaceuticals in the aquatic environment is recognized as one of the emerging issues in environmental chemistry. Conventional wastewater treatment plants (WWTPs) are not designed to remove pharmaceuticals (and their metabolites) from domestic wastewaters. The treatability of pharmaceutical compounds in WWTPs varies considerably depending on the type of compound since their biodegradability can differ significantly. As a consequence, they may reach the aquatic environment, directly or by leaching of the sludge produced by these facilities. Currently, the technologies under research for the removal of pharmaceuticals, namely membrane technologies and advanced oxidation processes, have high operation costs related to energy and chemical consumption. When chemical reactions are involved, other aspects to consider include the formation of harmful reaction by-products and the management of the toxic sludge produced. Research is needed in order to develop economic and sustainable treatment processes, such as bioremediation and biosorption. The use of low-cost materials, such as biological matrices (e.g., algae and fungi), has advantages such as low capital investment, easy operation, low operation costs, and the non-formation of degradation by-products. An extensive review of existing research on this subject is presented.

Keywords: algae; fungi; bioremediation; biosorption; removal of pharmaceuticals; wastewater

1. Introduction

The growth of world population, the promotion of health and better living conditions, and the rise of average life expectancy were accompanied by an increase in the consumption of pharmaceuticals, which are excreted in their original form or as metabolites and collected by the sewage system of urban wastewaters; they are not completely removed by wastewater treatment plants (WWTPs). Their occurrence in the aquatic environment is creating polluting pressure on aquatic ecosystems and is recognized as one of the emerging problems in the last decade [1]. The development of analytical techniques (such as gas chromatography mass spectrometry and liquid chromatography mass spectrometry) enabled the detection and quantification of a broad range of pharmaceuticals (and their metabolites) in environmental matrices [2]. Although they occur at extremely low levels (ranging from $\mu g \cdot L^{-1}$ to $ng \cdot L^{-1}$ or even lower), it is known that their presence in the environment is a potential hazard to public health [3–6].

The release of pharmaceuticals into the aquatic environment is not yet subject to regulation. In Europe, the "Water Framework Directive 2000/60/EC" introduced a strategy for water protection [7]. However, since the water situation in the several European Union (EU) countries was, and still is, different, it was necessary to proceed with adjustments to the original "Water Framework Directive".

In fact, one year after its publication, Decision 2455/2001/EC of the European Parliament and of the Council of 20 November 2001 established that each European Union member state should create, until 2009, a program of environmental protection measures for each of its hydrographic regions. A list of 33 priority substances in the field of water policy was also established, which includes pharmaceuticals. The revision and update of this list is done at least every four years, this may lead to the inclusion of new priority compounds, or removal, depending on the level of risk they pose [8]. Later, the "Directive on Environmental Quality Standards" (Directive 2008/105/EC) amended some environmental quality standards in the field of water policy and also established 11 new substances to be identified as priority substances or priority hazardous substances [9]. The Directive 2013/39/EU and Implementing Decision 2015/495 incorporated new substances, totalizing 45, implementing the first so-called watchlist. A few pharmaceuticals (diclofenac, erythromycin, clarithromycin, and azithromycin), as well as synthetic and natural hormones (estrone, 17β-estradiol, and 17α-ethinylestradiol), were included, and they should be carefully monitored by the Member States in order to set their environmental quality standards [10,11]. In its last review, five substances or groups of substances were removed, among them diclofenac, and three new substances were included, among them amoxicillin and ciprofloxacin. This is consistent with the European One Health Action Plan against Antimicrobial Resistance, which supports the use of the watchlist to improve knowledge and to evaluate the risks to human and animal health posed by the presence of antimicrobials in the environment. This review resulted in the publication of the second watchlist in 2018 (of the Commission Implementing Decision 2018/840 [12]. Also, in the United States, the Environmental Protection Agency (EPA) included 12 pharmaceuticals, personal care products, and endocrine disrupting compounds in a list in order to evaluate their related occurrence and safety risks. Therefore, is expected that, in the near future, legal limits will be established for the concentration of pharmaceuticals in WWTP discharges.

Indeed, pharmaceuticals are introduced into the environment (Figure 1) mainly through the discharge of treated effluents from conventional WWTPs, via domestic and hospital wastewaters, directly or by leaching of the sludge produced by these facilities, or through discharges of the pharmaceutical industry [13–15], as well as veterinary facilities where pharmaceuticals are widely used in livestock production for disease prevention and growth promotion [16–18].

The most frequently detected classes of pharmaceuticals in wastewaters are antibiotics, antiepileptics, antiphlogistics, X-ray contrast media, lipid-regulators, β-blockers, and tranquillizers [19]. Most of these compounds have high solubility, low hydrophobicity, and often negative charge at neutral pH (acidic compounds); these properties add more difficulties to their treatment [20]. WWTPs were not specifically designed to completely remove pharmaceuticals [21,22]. Removal efficiencies can vary from negligible to 100% depending on the compound [23]. The majority of compounds are removed in a percentage of less than 50%, which is related to the structure of the compounds [20]. Many physico-chemical and biological treatments were tested by WWTPs, but none of them are able to efficiently remove them [24]. Tertiary treatments, such as membrane technologies and advanced oxidation processes, usually play a more active role in the removal of these micropollutants than primary and secondary treatments [5,25]. However, they have many drawbacks in terms of energy requirement, large use of chemicals, and formation of undesired and harmful by-products [26]. To overcome these drawbacks, the scientific community focused its attention on the development of eco-friendly, economically viable and comparatively less expensive technologies [27]. Biological tertiary treatments of wastewaters are efficient, less expensive, and more eco-friendly than other technologies. The use of certain microorganisms gained importance in applied environmental microbiology. Fungi and algae-based treatments were pointed out as promising technologies for the remediation of pharmaceuticals. In this article, special attention is given to studies addressing pharmaceutical removal with fungi and algae via bioremediation and/or biosorption mechanisms (Figure 2).

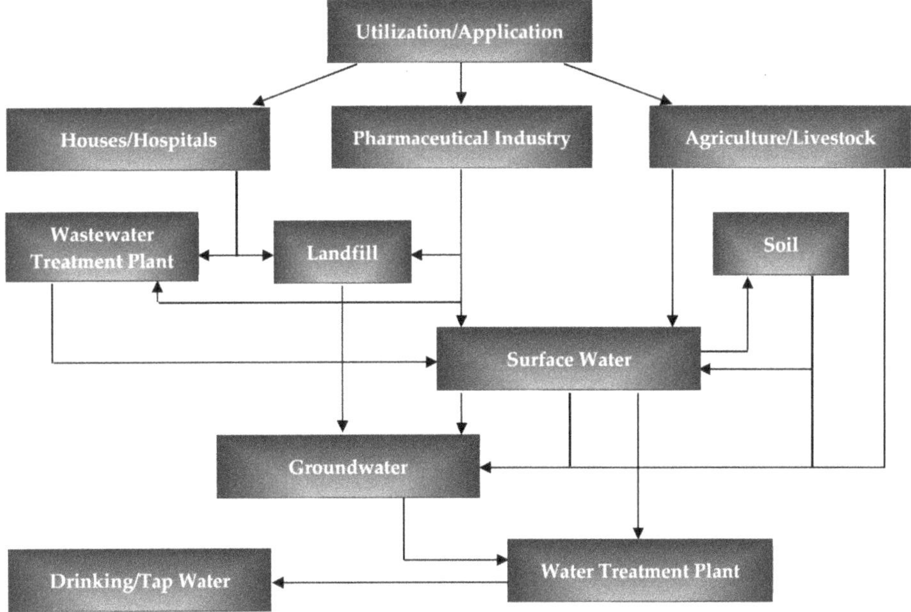

Figure 1. Pathways of pharmaceutical introduction into the environment.

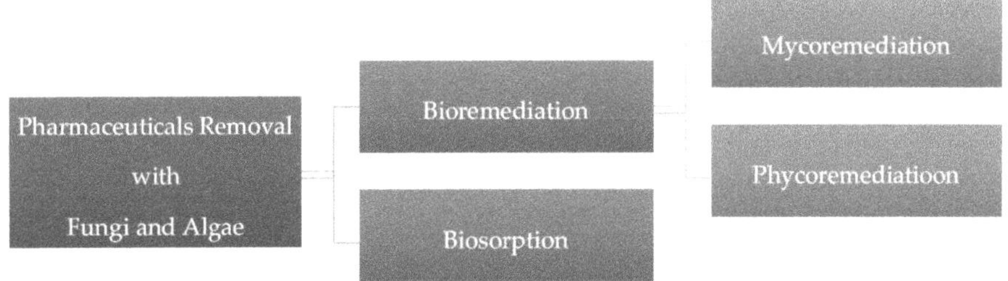

Figure 2. Mechanisms of pharmaceutical removal with fungi and algae.

2. Bioremediation

Bioremediation is a biological treatment that involves the use of microorganisms and their enzymes to convert recalcitrant and xenobiotic contaminants to less toxic forms and, therefore, short lifetimes in the environment or even their complete mineralization (the end-products are essentially carbon dioxide and water) [28]. Bioremediation of organic compounds was studied in more detail at the laboratory level, with the metabolic pathways of degradation known for some pharmaceutical compounds [29,30]. The process is influenced by environmental, physical, and chemical factors, namely, the stereochemistry, toxicity, and concentration of the contaminant, efficiency of the microbial strain, conditions during degradation (e.g., pH and temperature), retention time, presence of other compounds, and their concentration [31]. Some of bioremediation's advantages are as follows: it is accepted as a safe process, transforms pollutants instead of simply moving them from one medium to another [32], and presents lower costs when compared to the other technologies [33]. Although bioremediation proves to be a promising choice, research is needed to overcome some drawbacks of the process, which are the incomplete transformation, the limitation to biodegradable compounds, and the requirement of the selection and use of different microorganisms with specific metabolism for the different pollutants. Some of these disadvantages can be overcome through the use of genetically modified microorganisms [34].

2.1. Mycoremediation

Fungi ("mikes" from Greek) are eukaryotic organisms that include microorganisms such as molds, yeasts, and mushrooms. Some fungi are chemoheterotrophic organisms, being parasitic or saprophytic. Some are unicellular, and many are filamentous and have cell walls. The kingdom Fungi includes phyla Chytridiomycota (the chytrids), Zygomycota (the conjugated fungi), Ascomycota (the sac fungi), Basidiomycota (the club fungi), Deuteromycota (the imperfect fungi), and Glomeromycota. This classification was established according to their mode of sexual reproduction or using molecular data [35]. Fungi were efficiently used to treat water samples contaminated with micropollutants [36,37], pharmaceuticals in particular (Table 1). They are long recognized for their abilities to transform a broad range of recalcitrant compounds using nonspecific intracellular and extracellular oxidative enzymes [38–40]. The physiology and colonization strategy of mycelial fungi allows them to more easily withstand sudden changes in pH or humidity, as well as to degrade more efficiently complex organic compounds [41], although they are limited by a long growth cycle and spore formation [42].

2.1.1. Treatment Systems

Fungal reactors, also termed mycoreactors, can be suspended growth (such as slurry reactors) or immobilized systems (such as trickling filters, rotating biological contactors, upflow fixed-film reactors, and fluidized-bed reactors); these latter allow a fast biodegradation. Mycoreactors can be operated in batch, semi-batch, sequencing batch, or continuous mode, like other biological reactors. They can also be operated under aerobic or anaerobic conditions. Mycoreactors for submerged growth include stirred tanks, packed bed, bubble column, and air-lift [43], with stirred tanks as the most common, where the culture medium is agitated mechanically, providing a good fluid mixture and a good oxygenation. However, they have disadvantages such as the stress generated by the agitation, their impracticality for certain microorganisms, and the high energy consumption for high agitation speeds. Air-lifts are pneumatically agitated mycoreactors very similar to the stirred tanks; however, the fluid mixing is done by injecting air or compressed gas into the base of the bioreactor. This bioreactor is usually cylindrical so that the air bubbles remain as long as possible in the fluid. Compared with stirred tank, air-lifts have worse mixing rates; however, they generate less stress to the microorganisms [44,45]. Pellet-forming fungi, which are easier to recover at the end of the treatment, are cultivated in aerated fluidized-bed or suspended air-lift loop reactors [43].

2.1.2. Mechanisms of Removal

White-rot fungi (WRF), in particular, belong to the Basidiomycota phylum, whose potential was explored in several studies about the removal of pharmaceuticals (Table 1) [46–72]. These strains are filamentous wood-degrading fungi, ubiquitous in nature, able to mineralize lignin efficiently. The name white rot derives from the bleached appearance of the wood attacked by these fungi due to the removal of the dark-colored lignin. The same mechanism that gives these fungi the potential to degrade lignin also allows them to degrade a wide variety of recalcitrant pollutants, such as pharmaceuticals, making them promising and attractive microorganisms for wastewater bioremediation. They are able to mineralize a wide variety of pollutants since their enzymatic system is non-specific, non-stereoselective, and based on free-radical levels [73]. Pollutant degradation (Figure 3) seems to involve either an intracellular enzymatic system (i.e., cytochrome P450 system) or an extracellular enzymatic system (mainly lignin peroxidase, manganese peroxidase, laccase, and versatile peroxidase) [73,74]. Peroxidases are secreted during secondary metabolism of WRF in the presence of nitrogen, carbon, or sulfur limitations, while laccases are glycosylated multicopper oxidoreductases that are produced during primary metabolism [75]. Peroxidases generally show higher redox potential than laccases; however, peroxidases suffer deactivation in the presence of hydrogen peroxide while laccases do not [76–78]; laccase performance seems to be affected by the presence of chlorine ions forms [79]. Lignin peroxidase catalyzes the one-electron oxidation of various aromatic compounds, with subsequent formation of aryl cation radicals which are decomposed spontaneously by various pathways. Manganese peroxidase catalyzes the oxidation of Mn^{2+} to Mn^{3+}, which in turn can cause oxidation of several phenolic substrates [80]. Fungal laccases were reported in several studies since these enzymes have wide substrate ranges and use only oxygen as the final electron receptor, producing water as the only by-product [65,81–103]. In many cases, preference was given to the use of isolated enzymes (enzymatic bioremediation) instead of the use of fungi biomass in order to reduce the time of treatment, to avoid the lag phase of fungal growth, to reduce sludge production, and to facilitate process control [39]. It was found that *Phanerochaete chrysosporium*, a major WRF, does not have laccase genes [104]. *Trametes versicolor* is a WRF frequently cited in the literature that was shown to be effective in the removal of different pharmaceuticals (Table 1) [46,47,50–55,57–62,72].

For example, strains of *Trametes versicolor*, *Irpex lacteus*, *Ganoderma lucidum*, and *Phanerochaete chrysosporium* were simultaneously tested in the removal of ibuprofen, clofibric acid, and carbamazepine. The results suggested that clofibric acid and carbamazepine degradation occurred intracellularly by the cytochrome P450 system of *Trametes versicolor* [46]. Using the same fungus, Nguyen et al. [54] compared the removal of trace organic contaminants (including pharmaceuticals and steroid hormones) by alive, intracellular enzyme-inhibited and chemically inactivated whole-cell preparations, and a fungal extracellular enzyme extract, predominantly laccase. The low degradation of some hydrophobic compounds by the extracellular extract, and the impact of intracellular cytochrome P450 system inhibition on the degradation of some trace organic contaminants by the whole-cell culture indicated the importance of extracellular enzyme-independent catalytic pathways. Rodriguez-Rodriguez et al. [49] studied the removal of sulfonamides sulfapyridine and sulfathiazole by *Trametes versicolor*. Complete degradation was achieved for both compounds, although a longer period of time was needed to completely remove sulfathiazole when compared to sulfapyridine. In order to determine the effect of cytochrome P450 inhibitors, piperonyl butoxide or 1-aminobenzotriazole was added in the experiments performed. The results showed that sulfathiazole degradation was partially suppressed, while no additional effect was observed for sulfapyridine. In another study, *Trametes versicolor* was able to degrade carbamazepine in aqueous medium in an air-pulsed fluidized bioreactor in batch and continuous mode. In batch mode, carbamazepine concentration decreased 96%, while, in continuous mode, carbamazepine concentration decreased 54%. In this case, it was not possible to establish a correlation between extracellular laccase activity and carbamazepine degradation, since laccase and manganese peroxidase levels were negligible during the initial period, which may indicate that these enzymes were involved in an early stage of carbamazepine removal [47].

Rodarte-Moralez et al. [48] studied the removal of six pharmaceuticals (citalopram, sulfamethoxazole, diclofenac, ibuprofen, naproxen, and carbamazepine) from an initial mixture by three other WRF strains, *Bjerkandera* sp. R1, *Bjerkandera adusta*, and *Phanerochaete chrysosporium*, and also confirmed their enzymatic activity, particularly manganese peroxidase. An intense enzymatic activity was also detected during the individual degradation of 17α-ethinylestradiol and carbamazepine by ligninolytic fungi strains *Pleurotus* sp. P1, *Pleurotus ostreatus* BS, and a (unidentified) basidiomycete strain BNI. During 17α-ethinylestradiol degradation by *Pleurotus* sp. P1, laccase and manganese peroxidase activity was detected, while, during carbamazepine degradation by strain BNI, laccase, manganese peroxidase, and lignin peroxidase activity was detected [68].

Nguyen et al. [51,54] revealed in their studies that the enzymatic performance of laccase, in particular, can be enhanced by the addition of mediators. These authors studied the effect of continuous dosing of a mediator (1-hydroxybenzotriazole) in the removal of trace organic contaminants, including pharmaceuticals, from an initial complex mixture by *Trametes versicolor*. Becker et al. [63] studied the removal of 38 antibiotics (majority non-phenolic) from a mixture using immobilized laccase (*Trametes versicolor*), in an enzymatic membrane reactor, with or without the addition of syringaldehyde as a mediator. Thirty-two out of 38 antibiotics were removed by up to 50% when a mediator was used. In contrast, no significant removal was observed in experiments applied without a mediator.

Indeed, mediators can be oxidized by laccase to free radicals, which in turn can oxidize pollutants less specifically, increasing the variety of pollutants potentially degraded by these enzymes [105].

Synthetic and natural laccase mediators were used in enzymatic studies. Synthetic mediators include 2,2′-azino-bis(3-ethylbenzothiazoline-6sulfonate) and 1-hydroxybenzotriazole, while natural phenolic mediators include syringaldehyde and acetosyringone. Natural mediators are more economically feasible and more environmentally friendly than artificial mediators [106]. Each mediator has a specific catalytic mechanism [107,108]. The effect of a mediator depends on the radicals formed, the mediator recyclability, and the laccase stability in the mediator's presence [106,107,109,110]. Despite the proven increase in efficiency, mediators incur additional costs, and can cause toxicity [63,111] and laccase inactivation [112–114]. For example, Becker et al. [63] observed that, although the addition of syringaldehyde enhanced the removal of antibiotics, unspecific toxicity was also induced.

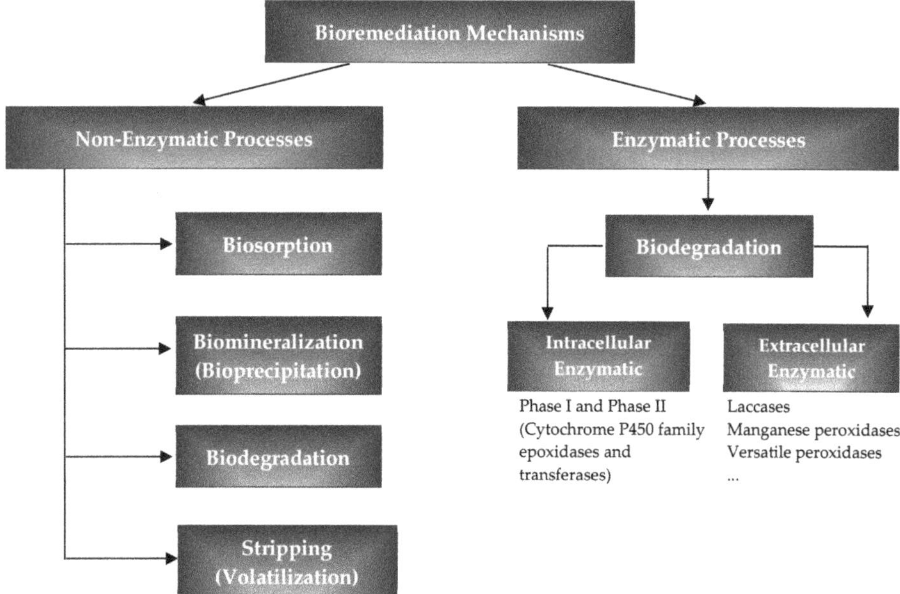

Figure 3. Fungi and algae strategies to counteract pharmaceuticals.

2.1.3. Factors That Influence the Degradation Capability

Fungi are able to degrade very low or non-detectable concentrations levels [73] and withstand a wide range of pH, further enhancing their degradation capability [115]. Nevertheless, mycoremediaton requires specific and controlled conditions in order to maintain a durable and efficient process. Fungi oxidative metabolism can be strongly affected by the presence of nutrients, pH, immobilization on different supports, and agitation/static growth conditions [31]. Zhang and Geißen [116] tested the degradation of carbamazepine and diclofenac by lignin peroxidase produced by WRF *Phanerochaete chrysosporium* in various conditions. It was found that lignin peroxidase completely degraded diclofenac at pH 3.0–4.5 and 3–24 mg·L^{-1} H_2O_2, while the degradation efficiency of carbamazepine was mostly below 10%. The addition of veratryl alcohol and the high temperature (30 °C) did not enhance the carbamazepine degradation. In another study, the removal of carbamazepine by *Phanerochaete chrysosporium* in a plate bioreactor operated in batch and continuous systems showed that carbamazepine removal depends on a sufficient nutrient supply. Carbamazepine concentration decreased about 80% when a diluted synthetic nutrient medium was fed to the reactor and decreased about 60% when a real effluent with additional glucose and nitrogen was fed to the reactor, after 100 days of incubation [49].

By opposition, 17α-ethinylestradiol was completely degraded by *Pleurotus* sp. P1, *Pleurotus ostreatus* BS, and a (unidentified) basidiomycete strain BNI, in the presence or absence of another carbon source [68].

Becker et al. [57] showed that the degradation of hormones by laccases is feasible even at very low enzyme concentrations, and that immobilized enzymes displayed better removal performance compared to the free enzyme. Indeed, enzyme immobilization provides a more suitable environment for enzymes and may result in an increasing enzyme stability concerning pH, temperature, and storage time. Efficient use of immobilized cells requires, however, control of physiological/metabolic changes occurring either during the immobilization process or during the biotransformation phase. Enzyme immobilization can be done with irreversible methods (e.g., covalent binding and entrapment) or reversible methods (e.g., adsorption, ionic binding, affinity binding, chelation, and disulfide bonds). The selection of the immobilization method is based on technical–economic criteria. Since cost is a preponderant parameter, the simplest ones are usually used [117].

An eventual contamination with bacteria can negatively affect the efficiency since it can generate competition for the substrate, damage fungi mycelium, disrupt biomass growth, and destabilize fungal activity [118,119]. To avoid bacterial contamination, some strategies can be implemented. In the biological treatment system, the pH has great influence, since it acts in the development and selection of the microorganisms. Therefore, it is convenient to operate under acidic conditions, since the optimum pH for fungi growth is lower than the optimum pH for most bacteria. Other possible strategies include pretreatment by coagulation–flocculation of wastewater, since a reduction of the initial bacterial count usually occurs; coupling the bioreactor with a micro-screen, which allows the retention of fungal biomass and, simultaneously, the washout of bacteria with effluent; the use of disinfectant agents that allow the selection and/or inactivation of bacteria; the immobilization of fungal strains in different carriers; and the periodic replacement of fungal biomass [118].

Sterility conditions do not appear to be a mandatory requirement to apply *Trametes versicolor* in pharmaceutical removal. The studies performed showed that this microorganism was able to partially or completely remove pharmaceuticals either under sterile or non-sterile conditions [50,52].

2.1.4. Concluding Remarks and Future Challenges for Mycoremediation

In general, studies indicate a significant reduction in wastewater toxicity after treatment. Some fungi strains found in the literature [46,47,50,52,53,61,120] proved to be effective in pharmaceutical removal (Table 1). Final products were reported to be less toxic or more biodegradable than the parent compounds, which emphasizes the potential of fungi as remedial agents.

Genetic tools can represent an essential step to improve fungi performance and overcome some limitations associated with the process. The sequencing of fungi genomes will allow the development of new genetic techniques to improve metabolic and adaptive processes, and consequently achieve efficient bioremediation [121]. To date, efforts met limited success, and there is still a long way to go before the introduction of modified fungi in remediation processes. The first complete eukaryotic genome belongs to the yeast *Saccharomyces cerevisiae* [122]. Several projects released information about the genome sequences of fungi *Aspergillus nidulans*, *Aspergillus fumigatus*, *Neurospora crassa*, and *Coprinus cinereus* [123]. The thirty-million-base-pair genome of WRF *Phanerochaete chrysosporium* strain RP78 was sequenced using a whole-genome shotgun approach [124]. The genome revealed genes encoding oxidases, peroxidases, and hydrolytic enzymes involved in wood decay, which opens new horizons related to the process of biodegradation of organic pollutants and pharmaceuticals in particular [125]. However, the risks involved in the use of genetically modified fungi, their impact on the environment and human health, and existing legal limitations must be considered [126].

2.2. Phycoremediation

Algae ("phyco" from Greek) include prokaryotic organisms (cyanobacteria) and eukaryotic organisms (all the algae species) that contain chlorophyll and carry out oxygenic photosynthesis. Although most algae are of microscopic size and, hence, are clearly microorganisms (microalgae), several forms are macroscopic (macroalgae) grow to over 30 m in length. Algae are either unicellular or colonial. When the cells are arranged end to end, the alga is said to be in filamentous form [127]. According to Ruggiero et al. [128], algae can be classified according to the pigments they possess. Algae color differences arise due to the proportions of different auxiliary photosynthetic pigments present in addition to the green chlorophylls. Chlorophyta (the green algae) and Euglenophyta (the euglenoids) are green since chlorophyll *a* is dominant. If carotenoids are dominant in the algae, they give them a golden-brown color, such as Chrysophyta, whose chloroplasts contain chlorophylls *a*, c_1, and c_2, fucoxanthin, and β-carotene, which are the carotenoids responsible for the golden-brown color. Dinophyta (the dinoflagellates), have a reddish, greenish, or brown appearance due to chlorophylls *a* and c_2 and carotenoids. Rhodophyta (the red algae) have chlorophylls *a* and *d*, phycobiliproteins, and floridean starch as storage products accumulated in the cytoplasm outside the chloroplast. In Phaeophyta (the brown algae), color results from the dominance of fucoxanthin over chlorophylls *a*, c_1, and c_2.

Algae are highly adaptive microorganisms and can grow autotrophically, heterotrophically, or mixotrophically. They can grow in very harsh environmental conditions, such as low nutrient levels, and extreme pH and temperature, which is an advantage over some species of fungi [129]. Unlike strictly heterotrophic microorganisms, the decrease in nutrient concentration does not limit the growth of algae [130]. Microalgae can acclimatize to changes in temperature, salinity, light, and nutrient availability, which allows the improvement of their tolerance and biodegradation capacity. This adaptation mechanism to extreme conditions is explained by genetic changes caused by spontaneous mutation or physiological adaptation [131,132]. Xiong et al. [133] evaluated the biodegradation capability of *Chlorella vulgaris* after acclimation with multiple exposures to levofloxacin and an increase in salinity. Results showed that levofloxacin biodegradation was significantly improved after acclimation.

The characteristics of domestic wastewaters is usually suitable for the cultivation of microalgae, since wastewaters represent a source of nutrients. From an environmental perspective, photosynthetic microalgae are fascinating since they can sequester atmospheric carbon dioxide for their own growth, contributing to the mitigation of this pollutant. In a properly controlled process, the association of microalgae with bacteria may result in a very synergistic relationship; microalgae provide oxygen while bacteria release carbon dioxide, which allows a significant reduction of oxygen needs in the treatment process [134–137]. In addition, microalgae treatment supplies an environment that increases the mortality of pathogenic organisms due to the pH elevation [138]. A win–win situation of using microalgae in wastewater bioremediation offers a tertiary biotreatment of wastewater coupled with the production of potentially valuable biomass as a bioresource for biofuel or high-value by-products [139].

2.2.1. Treatment Systems

From the economic point of view, open systems are preferred for wastewater phycoremediation. Operational factors that influence algae growth are essentially mixing, dilution rate, and depth [140]. Open ponds are the most usual for microalgae cultivation since they require lower investment costs and operational capital. Commercially, algae cultivation is mainly performed on open channels stirred with a paddlewheel (raceway), since they are less expensive and easier to construct and operate. There is, however, a high risk of contamination and low productivity due to a poor mixing regime and light penetration, and also due to the difficulty of controlling the operating conditions [141]. Facultative, maturation, and high-rate algal ponds are the most used open systems for wastewater treatment, with the first two being the most used. The major differences between them are the depth and origin of the influent. The major constraints include poor light utilization by the cells, water loss due to evaporation, diffusion of carbon dioxide to the atmosphere, temperature fluctuations, inefficient stirring, and large space requirements for microorganism growth [142]. Open systems usually operate under long hydraulic retention time (between eight and 14 days) in order to consume carbon dioxide during the day (photosynthesis) and provide oxygen for aerobic biodegradation. Sunlight intensity influences photosynthetic activity, leading to pH and dissolved oxygen variations [143–145].

Closed systems were designed to overcome the problems associated with open systems. Unlike open systems, closed systems allow greater control of the process; however, they are expensive to install and to maintain [146]. Closed systems are most suitable for pure algae strains, and their design must be carefully optimized for each individual strain according to its unique physiological and growth characteristics. These systems avoid losses by evaporation, and reduce the contamination risk and the losses of carbon dioxide to the atmosphere [147]. Closed systems, which are mostly photobioreactors, exist in various design configurations (e.g., horizontal or serpentine tube, flat-plate, bubble column, air-lift column, and stirred tank). Flat-plate photobioreactors have space constraints, and tubular photobioreactors have design limitations. Column photobioreactors are the most efficient, which provide efficient mixing, the highest volumetric mass transfer rates, and the best controllable growth conditions. Some difficulties arise with scale-up caused by inhomogeneous distribution of light inside the culture. The productivity is negatively affected by central, light-deprived zones [148]; therefore, to facilitate penetration of light, photobioreactors are made of glass, transparent plastic material, or sturdy polythene.

Among all the drawbacks associated with phycoremediation, harvesting is one of the most challenging processes integrated in the removal treatment. The harvesting process may account for 30% of total microalgae biomass production cost. Several factors affect the harvesting, such as microalgae strain and operational parameters (hydraulic and solid retention times) [149]. Harvesting technologies may involve one or more steps, and different physical (e.g., centrifugation, gravity sedimentation, filtration, and dissolved air flotation), chemical (e.g., chemoflocculation), and biological (e.g., bioflocculation and microalgae immobilization) processes.

Most of these processes are limited by high costs, long processing times, high energy consumption, and low recovery [150,151]. The choice of the harvesting method depends on the microalgae characteristics, such as the density and size of algal cells, as well as the product derived from the algal biomass [152].

The most environmentally friendly method is bioflocculation as it involves formation of extracellular biopolymers that help natural flocculation of small particles [150,151]. Some microalgae species flocculate more readily than others, and these microalgae can be mixed with other species to induce flocculation [153,154]. However, this method may be not efficient due to the small size of microalgae cells and their fast growth. Further research is required to understand the underlying mechanism associated with this process.

New technologies were considered, namely, the addition of other microorganisms, such as bacteria and some fungal species, to the microalgae culture [135]. In the first case, microalgae and bacteria may form flocs that settle more easily than single microalgae. It is necessary to add an extra organic substrate to allow bacteria growth [135], which can be provided by using wastewater. The presence of an organic carbon source in wastewater allows both organisms to thrive together [155,156]. The main disadvantages are the influence of dominant microalgae species and the variability of wastewaters [135]. In the second case, some filamentous fungi can pelletize, entrapping the microalgae cells, which facilitates the harvest by simple sieve filtration or sedimentation in most cases [157]. It is not necessary to add an extra sugar to allow fungi growth [158]. Co-pelletization efficiency seems to depend both on fungi and microalgae strains and on culture conditions (in particular on the pH value). The choice of fungi strains will be the key issue since it determines the overall pelletization efficiency and, therefore, can have a direct impact on the subsequent processes. Since this method does not require the addition of chemicals or inputs of energy, it may offer a solution to two of the major problems of harvesting processes, the high cost and high energy consumption [159]. The competition of nutrients between fungi and microalgae is believed to occur mostly in heterotrophic conditions, since microalgae need external carbon sources to support their growth, or in autotrophic conditions, when other nutrients, such as organic nitrogen, are limited. In this context, neither fungi nor microalgae cells will reach their maximum growth compared with their growth in individual pure cultures [160]. It is not clear if co-culturing fungi affects the growth and nutrient assimilation of microalgae [161]. It is also unclear how the use of fungi for pelletization is fundamentally different or superior to the use of bacteria for bioflocculation [161,162].

2.2.2. Mechanisms of Removal

Studies about the removal of emerging contaminants (such as pharmaceuticals) with algae are limited; therefore, the mechanisms involved are not yet very clear (Table 1) [163]. The possible mechanisms involved in algal–bacterial systems (Figure 3) seems to be biodegradation, biomineralization (bioprecipitation), biosorption (cell adsorption and/or bioaccumulation), stripping (volatilization), and photodegradation, due to the effect of light [163–165]. However, volatilization can be considered negligible for most pharmaceuticals, because of their low Henry's constant values [166]. The bioaccumulation of pharmaceuticals in algae cells can induce the generation of reactive oxygen species, free radicals (e.g., $O_2^{\bullet}$—superoxide radicals, $OH^{\bullet}$—hydroxyl radical, $HO_2^{\bullet}$—perhydroxy radical, and $RO^{\bullet}$—alkoxy radicals) and nonradical forms (e.g., H_2O_2—hydrogen peroxide and $^1O_2$—singlet oxygen). At normal levels, these species act as essential signaling molecules to control cellular metabolism; however, at excess levels, these species can cause severe damage to cellular components and an increased rate of mutagenesis that ultimately leads to programmed cell death [167]. Such as for fungi, pollutant degradation by algae seems to involve intracellular and extracellular enzymatic systems (Figure 3). Intracellular degradation of pharmaceuticals involves a phase I enzyme (cytochrome P450) [163,168–173] and a phase II enzyme (e.g., glutathione-*S*-transferases) [171]. Extracellular degradation of pharmaceuticals involves the excretion of various extracellular polymeric substances, such as polysaccharides, protein, enzymes, substituents (polysaccharide-link methyl and acetyl groups), and lipids to their surrounding environment. These extracellular polymeric substances can form a hydrated biofilm matrix that acts as an external digestive system since they keep extracellular enzymes close to the cells [174].

Peng et al. [168] studied the removal of progesterone and norgestrel by the microalgae *Scenedesmus obliquus* and *Chlorella pyrenoidosa*. According to the authors, biotransformation was found to be the main mechanism for the removal of progestogens. Hydroxylation, oxidation/reduction, and side-chain breakdown were proposed to be involved in the algal transformation of the target compounds by *Chlorella pyrenoidosa*. Xiong et al. [169] studied the removal of carbamazepine by the microalgae *Chlamydomonas mexicana* and *Scenedesmus obliquus*. The results showed that both species simultaneously promoted biodegradation, adsorption, and bioaccumulation of carbamazepine. De Godos et al. [164] studied the mechanisms of tetracycline removal from a synthetic wastewater by *Chlorella vulgaris* and identified photodegradation and biosorption as the most important.

2.2.3. Factors That Influence the Degradation Capability

There are several factors that affect the degradation of pharmaceuticals, which are mainly related to the operation conditions, reactor configuration, and the species present. Matamoros et al. [175] studied the removal of emerging organic compounds, including pharmaceuticals, from a real wastewater by the marine algae *Lessonia nigrescens* Bory and *Macrocystis integrifolia* Bory in two pilot-scale high-rate algal ponds. Removal efficiency ranged from negligible to up to 90% and was only affected by the hydraulic retention time during the cold season. This effect was not observed in the warm season.

De Godos et al. [164] studied the removal of tetracycline by *Chlorella vulgaris* in two pilot-scale high-rate algal ponds operated in batch mode. The study demonstrated that the shallow geometry of high-rate algal ponds is advantageous to support tetracycline's photodegradation.

Díaz-Garduño et al. [176] studied the removal of organic compounds, including pharmaceuticals, from a real wastewater, by the microalga *Coelastrum* sp., in a pilot-scale photobioreactor and a multibarrier treatment. The multibarrier treatment was the most effective treatment regarding the removal efficiencies. The photobioreactor showed different removal percentages depending on the initial effluent composition.

Hom-Diaz et al. [177] observed that the removal of pharmaceuticals from domestic wastewater in a pilot-scale tubular photobioreactor, in two seasonal periods (September–October; October–December), was highly impacted by temperature and solar irradiation.

The removal of several pharmaceuticals present in wastewater by green algal species, in a real-scale photobioreactor, was positively correlated with light intensity inside the culture, with stronger correlation when the data collected during the night were excluded [178].

Lai et al. [179] studied the removal of natural steroid estrogens (estradiol, estrone, estriol, and hydroxyestrone) and synthetic steroid estrogens (estradiol valerate, estradiol, and ethinylestradiol) by the alga *Chlorella vulgaris*, using batch-shaking experiments in the light and in the dark. The results showed that estradiol and estrone were interconvertible in the presence or absence of light. In the presence of light, 50% of estradiol was further metabolized to an unknown product. Estradiol valerate was hydrolyzed to estradiol and then to estrone. Estrone, hydroxyestrone, estriol and ethinylestradiol were relatively stable, and did not suffer biotransformation. Recently, a combination of photobioreactor and open-pond cultivation was suggested; the first allows a fast algae growth, while the second ensures mass cultivation [141].

In many cases, the complete removal of a compound requires the interaction of several groups of microorganisms (consortium), each of which is responsible for a degradation step. This interaction is extremely positive because pure cultures, isolated from this consortium, may not be able to completely remove this compound as the single carbon source, or the removal rate may be significantly lower than that obtained with the mixed crop that gave rise to it. Shi et al. [180] studied the removal of synthetic hormone 17α-ethinylestradiol and the natural hormones estrone and 17β-estradiol, from a synthetic wastewater, by an alga (*Anabaena cylindrica*, *Chlorococcus*, *Spirulina platensis*, *Chlorella*, *Scenedesmus quadricauda*, and *Anaebena*) and duckweed (*Lemna*) pond system. The results showed that all hormones were effectively removed from the continuous-flow algae and duckweed pond even when their concentrations were at ng·L^{-1} level. The simultaneous presence of algae and duckweed accelerated the removal of hormones from the synthetic wastewater since hormones could be quickly sorbed either on duckweed or algae and then degraded by both microorganisms [180].

Although several studies demonstrated the applicability of algae for micropollutant removal from wastewater [137,181], limitations and knowledge gaps still exist to rely on algae biomass production as an effective mean for micropollutant removal. For example, when Wang et al. [182] exposed the freshwater alga *Chlorella Pyrenoidosa* to a triclosan concentration of 800 mg·L^{-1}, a reductive dechlorination product of triclosan was formed. Algal cell growth was affected by their toxicity. Algal cell chloroplasts were damaged, decreasing the energy supply for algal growth, which produced an adverse effect on the effectiveness of triclosan biodegradation. Furthermore, most of the studies tested the algae's ability to remove pollutants, which grew in unpolluted media before the treatment, and ignored the toxic stress caused by the pollutant and its influence on the removal capability. It is possible that the sensitivity or tolerance of algae changes after the first contact with the pollutant and, therefore, it may influence the removal efficiency of subsequent batch treatments.

Chen et al. [183] investigated the removal efficiency of cefradine by *Chlorella pyrenoidosa* in a sequencing batch reactor and identified different results between the first batch treatment and the second, where higher algal growth inhibition rates were observed; however, the alga produced more photosynthetic pigments, enhancing its photosynthetic metabolism as a way of adaptation to more harmful environmental conditions.

2.2.4. Concluding Remarks and Future Challenges

Some algae strains found in the literature [64,133,148,164,166,168–170,176–180,182–194] proved to be effective in pharmaceutical removal (Table 1).

Table 1. Bioremediation studies with fungi and algae for removal of pharmaceuticals. WWTP—wastewater treatment plant.

Compounds	Compounds Source	Strains	Removal Mechanisms	Technologies	Reference
Ibuprofen, clofibric acid, and carbamazepine	Synthetic media	Fungi *Trametes versicolor* *Irpex lacteus* *Ganoderma lucidum* *Phanerochaete chrysosporium*	Biodegradation Adsorption	Laboratory-scale batch assays	[46]
Carbamazepine	Synthetic media	Fungi *Trametes versicolor*	Biodegradation Adsorption	Laboratory-scale batch assays Pilot-scale glass air-pulsed fluidized bioreactor (continuous and batch feed)	[47]
Citalopram, fluoxetine, sulfamethoxazole, diclofenac, ibuprofen, naproxen, carbamazepine, and diazepam	Synthetic media	Fungi *Bjerkandera* sp. R1 *Bjerkandera adusta* *Phanerochaete chrysosporium*	Biodegradation	Laboratory-scale batch assays	[48]
Carbamazepine	Synthetic media	Fungi *Phanerochaete chrysosporium*	Biodegradation Biosorption	Pilot-scale plate bioreactor (continuous and batch feed)	[49]
Naproxen, ibuprofen, acetaminophen, salicylic acid, ketoprofen, codeine, erythromycin, metronidazole, ciprofloxacine, azithromycin, cefalexine, propranolol, carbamazepine, 10,11-epoxycarbamazepine, 2-hydroxycarbamazepine, acridone, and citalopram	Urban wastewater	Fungi *Trametes versicolor*	Biodegradation	Pilot-scale air-fluidized bioreactor (batch feed)	[50]
Carbamazepine, ibuprofen, clofibric acid, ketoprofen, metronidazole, triclosan, 17-α-ethinylestradiol, 17-β-estradiol-17-acetate, estrone, estriol, 17-β-estradiol, gemfibrozil, amitriptyline, primidone, salicylic acid, diclofenac, naproxen	Synthetic media	Fungi *Trametes versicolor*	Biodegradation Biosorption	Laboratory-scale batch assays Pilot-scale fungus-augmented membrane bioreactor (continuous feed)	[51]
Acetaminophen, ibuprofen, ketoprofen, naproxen, salicylic acid, codeine, phenazone, dexamethasone, diclofenac, piroxicam	Hospital wastewater	Fungi *Trametes versicolor*	Biodegradation	Pilot-scale glass air pulsed fluidized bioreactor (batch feed)	[52]
X-ray contrast agent iopromide and antibiotic ofloxacin	Hospital wastewater	Fungi *Trametes versicolor*	Biodegradation	Laboratory-scale batch assays Pilot-scale glass air-pulsed fluidized bioreactor (batch feed)	[53]
Metronidazole, salicylic acid, primidone, amitriptyline, carbamazepine, ketoprofen, naproxen, ibuprofen, gemfibrozil, diclofenac, triclosan, estriol, estrone, 17-α-ethinylestradiol, 17-β-estradiol, 17-β-estradiol-17-acetate	Synthetic media	Fungi *Trametes versicolor*	Biodegradation Biosorption	Laboratory-scale batch assays	[54]
Cefalexin, ciprofloxacin, metronidazole, trimethoprim, tetracycline, ketoprofen, acridone, carbamazepine, a carbamazepine metabolite, ciprofloxacin, metronidazole and its hydroxilated metabolite, β-blocker carazolol, diazepam, naproxen, cephalexin, tetracyclin, sertraline, paroxetine, gemfibrozil, amlodipine, furosemide, dimetridazole, azythromycin, ronidazole, olanzapine, piroxicam, β-blockers metoprolol	Veterinary hospital wastewater	Fungi *Trametes versicolor*	Biodegradation	Pilot-scale glass air-pulsed fluidized bioreactor (continuous and batch feed)	[55]

Table 1. *Cont.*

Compounds	Compounds Source	Strains	Removal Mechanisms	Technologies	Reference
Acetaminophen, carbamazepine, diclofenac, metoprolol, naproxean, ranitidine, and sulfamethoxazole	Synthetic media	Fungi *Aspergillus niger* Algae *Chlorella vulgaris*	Biodegradation	Laboratory-scale batch assays	[56]
Estrone, 17β-estradiol, 17α-ethinyl-estradiol, and estriol	WWTP wastewater	Fungi *Trametes versicolor* *Myceliophthora thermophila*	Biodegradation Adsorption	Laboratory-scale batch assays	[57]
17β-estradiol and 17α-ethinylestradiol	Synthetic media	Fungi *Trametes versicolor*	Biodegradation	Laboratory-scale batch assays Pilot-scale glass air-fluidized bioreactor (continuous feed)	[58]
Sulfapyridine, sulfapyridine, and sulfamethazine	Synthetic media	Fungi *Trametes versicolor*	Biodegradation Biosorption	Laboratory-scale batch assays Pilot-scale air-pulsed fluidized-bed bioreactor (continuous feed)	[59]
Naproxen and carbamazepine	Synthetic media	Fungi *Trametes versicolor*	Biodegradation	Laboratory-scale batch assays	[60]
Sodium diclofenac	Synthetic media	Fungi *Trametes versicolor*	Biodegradation	Laboratory-scale batch assays	[61]
Ketoprofen	Synthetic media	Fungi *Trametes versicolor*	Biodegradation	Laboratory-scale batch assays	[62]
Diclofenac, ibuprofen, naproxen, carbamazepine, and diazepam	Synthetic media	Fungi *Phanerochaete chrysosporium*	Biodegradation Adsorption	Pilot-scale stirred tank reactor and fixed-bed reactor (continuous feed)	[64]
Tetracycline and oxytetracycline	Synthetic media	Fungi *Phanerochaete chrysosporium*	Biodegradation	Laboratory-scale batch assays	[65]
Phenolic compounds	Pharmaceutical industry wastewater	*Pycnoporus sanguineus*	Biodegradation	Laboratory-scale batch assays	[66]
Diclofenac, ketoprofen and atenolol	Hospital wastewater	Fungi *Pleurotus ostreatus*	Biodegradation Adsorption	Pilot-scale air-pulsed fluidized-bed bioreactor (continuous and batch feed) Laboratory-scale batch assays	[67]

Table 1. *Cont.*

Compounds	Compounds Source	Strains	Removal Mechanisms	Technologies	Reference
17α-ethinylestradiol and carbamazepine	Synthetic media	Fungi *Pleurotus* sp. P1 *Pleurotus ostreatus* BS (unidentified) basidiomycete strain BNI	Biodegradation Adsorption	Laboratory-scale batch assays	[68]
Acetaminophen	Synthetic media	Fungi *Mucor hiemalis*	Bioconcentration	Laboratory-scale batch assays	[69]
Carbamazepine and clarithromycin	Synthetic media	Fungi *Trichoderma harzianumPleurotus ostreatus*	Biodegradation	Laboratory-scale batch assays	[70]
Carbamazepine	Synthetic media	Fungi *Pleurotus ostreatus*	Biodegradation	Laboratory-scale batch assays	[71]
Clofibric acid, gemfibrozil, ibuprofen, fenoprofen, ketoprofen, naproxen, diclofenac, indomethacin, propyphenazone, and carbamazepine	Synthetic media	Fungi *Trametes versicolor*	Biodegradation	Laboratory-scale batch assays	[72]
Levofloxacin	Synthetic media	Algae *Chlorella vulgaris*	Biodegradation Bioaccumulation	Laboratory-scale batch assays	[133]
Acetaminophen, ibuprofen, ketoprofen, naproxen, carbamazepine, diclofenac, and triclosan	WWTP wastewater	Algae *Lessonia nigrescens Bory* *Macrocystis integrifolia Bory*	Biodegradation Photodegradation Biosorption	Pilot-scale high-rate algal ponds	[148]
Tetracycline	Synthetic media	Algae *Chlorella vulgaris*	Photodegradation Biosorption	Laboratory-scale batch assays Pilot-scale high rate algal ponds (batch feed)	[164]
Carbamazepine	Synthetic media	Algae *Pseudokirchneriella subcapitata* (and crustacean *Thamnocephalus* and cnidarian *Hydra attenuata*)	Bioaccumulation	Laboratory-scale batch assays	[166]
Progesterone and norgestrel	Synthetic media	Algae *Scenedesmus obliquus* *Chlorella pyrenoidosa*	Biodegradation Biosorption	Laboratory-scale batch assays	[168]
Carbamazepine	Synthetic media	Algae *Chlamydomonas mexicana* *Scenedesmus obliquus*	Biodegradation Adsorption Bioaccumulation	Laboratory-scale batch assays	[169]

Table 1. *Cont.*

Compounds	Compounds Source	Strains	Removal Mechanisms	Technologies	Reference
17α-Ethynylestradiol	Synthetic media	Algae *Desmodesmus subspicatus*	Biotransformation Bioconcentration	Laboratory-scale batch assays	[170]
Analgesic and antiinflammatories, lipid regulators and antihypertensive, psychiatric drugs and stimulant, antibiotics, and others	WWTP wastewater	Algae *Coelastrum* sp.	Biodegradation	Pilot-scale photobiotreatment microalgae and multi-barrier treatment	[176]
Acetaminophen, ibuprofen, naproxen, salicylic acid, ketoprofen, codeine, azithromycin, erythromycin, ciprofloxacin, ofloxacin, atenolol, lorazepam, alprazolam, paroxetine, hydrochlorothiazide, furosemide, and diltiazem	Domestic wastewater	Algae Undefined microalgae	Biodegradation	Pilot-scale tubular photobioreactor	[177]
Alfuzosin, alprazolam, atenolol, atracurium, azelastine, biperiden, bisoprolol, bupropion, carbamazepin, cilazapril, ciprofloxacin, citalopram, clarithromycine, clemastine, clindamycine, clonazepam, clotrimazol, codeine, cyproheptadine, desloratidin, dicycloverin, diltiazem, diphenhydramin, eprosartan, fexofenadine, flecainide, fluconazole, flupetixol, haloperidol, hydroxyzine, ibersartan, loperamide, memantin, metoprolol, miconazole, mirtazapine, nefazodon, orphenadrin, pizotifen, ranitidine, risperidone, roxithromycine, sertraline, sotalol, sulfamethoxazol, terbutalin, tramadol, trihexyphenidyl, trimetoprim, venlavafaxin, and verapamil	WWTP wastewater	Algae Green algal species (*Tetradesmus dimorphus* and *Dictyosphaerium*, between them)	Biodegradation	Real-scale photobioreactor	[178]
Estradiol, estrone, estriol and hydroxyestrone) and synthetic steroid estrogens (estradiol valerate, estradiol, and ethinylestradiol	Synthetic media	Algae *Chlorella vulgaris*	Biotransformation Bioconcentration	Laboratory-scale batch assays	[179]
17α-ethinylestradiol, estrone, and 17β-estradiol	Synthetic media	Algae *Anabaena cylindrical* *Chlorococcus* *Spirulina platensis* *Chlorella* *Scenedesmus quadricauda* *Anaebena* (and duckweed *Lemna*)	Biodegradation Adsorption	Laboratory-scale batch assays Pilot-scale plug flow reactor (continuous feed)	[180]
Triclosan	Synthetic media	Algae *Chlorella Pyrenoidosa*	Biodegradation Biosorption	Laboratory-scale batch assays	[182]
Cefradinegree	Synthetic media	Algae *Chlorella pyrenoidosa*	Biodegradation	Pilot-scale batch-sequencing reactor algae process (batch feed)	[183]

Table 1. *Cont.*

Compounds	Compounds Source	Strains	Removal Mechanisms	Technologies	Reference
Ethinylestradiol	Synthetic media	Algae *Ankistrodesmus braunii* *Chlorella ellipsoidea* *Chlorella pyrenoidosa* *Chlorella vulgaris* *Scenedesmus communis* *Scenedesmus obliquus* *Scenedesmus quadricauda* *Scenedesmus vacuolatus* *Selenastrum capricornutum*	Biotransformation	Laboratory-scale batch assays	[184]
17α-boldenone, 17β-boldenone, 4-hydroxy-androst-4-ene-17-dione, androsta-1,4-diene-3,17-dione, 4-androstene-3,17-dione, carbamazepine, ciprofloxacin, clarithromycin, climbazole, clofibric acid, diclofenac, enrofloxacin, erythromycin–H_2O, estrone, fluconazole, gemfibrozil, ibuprofen, lincomycin, lomefloxacin, norfloxacin, ofloxacin, paracetamol, progesterone, roxithromycin, salicylic acid, salinomycin, sulfadiazine, sulfadimethoxine, sulfameter, sulfamethazine, sulfamethoxzole, sulfamonomethoxine, sulfapyridine, testosterone, triclosan, trimethoprim, and tylosin	WWTP wastewater	Algae *Chlamydomonas reinhardtii* *Scenedesmus obliquus* *Chlorella pyrenoidosa* *Chlorella vulgaris*	Biodegradation	Laboratory-scale batch assays	[185]
Salicylic acid and paracetamol	Synthetic media	Algae *Chlorella sorokiniana*	Biodegradation	Pilot-scale reactor (batch and semicontinuous feed)	[186]
Paracetamol, salicylic acid, and diclofenac	Synthetic media	Algae *Chlorella sorokiniana* *Chlorella vulgaris* *Scenedesmus obliquus*	Biodegradation	Pilot-scale bubbling column photobioreactor (batch and semicontinuous feed)	[187]
Tributyltin	Synthetic media	Algae *Chlorella miniata* *C. sorokiniana* *Scenedesmus dimorphus* *S. platydiscus*	Biodegradation Adsorption Absorption	Laboratory-scale batch assays	[188]
Trimethoprim, sulfamethoxazole, and triclosan	Synthetic media	Algae *Nannochloris* Sp.	Biodegradation Adsorption	Laboratory-scale batch assays	[189]
Diclofenac, ibuprofen, paracetamol, metoprolol, carbamazepine and trimethoprim, estrone, 17β-estradiol, and ethinylestradiol	Synthetic media	Algae *Chlorella sorokiniana*	Biodegradation Photolysis Biosorption	Laboratory-scale batch assays	[190]

Table 1. *Cont.*

Compounds	Compounds Source	Strains	Removal Mechanisms	Technologies	Reference
Oxytetracycline, doxycycline, chlortetracycline, and tetracycline	Synthetic media	Algae *Haematoloccus pluvialis* *Chlorella* sp. *Selenastrum capricornutum* *Pseudokirchneriella subcapitata*	Biodegradation	Laboratory-scale batch assays	[191]
Salicylic acid or paracetamol	Pharmaceutical industry wastewater	Algae *Chlorella sorokiniana*	Biodegradation	Pilot-scale bubbling column photobioreactor (batch and semicontinuous feed)	[193]
Ciprofloxacin	Synthetic media	Algae *Chlamydomonas mexicana*	Biodegradation	Laboratory-scale batch assays	[194]
β-estradiol and 17α-ethinylestradiol	WWTP anaerobic sludge	Algae *Selenastrum capricornutum* *Chlamydomonas reinhardtii*	Biodegradation Photodegradation Adsorption	Laboratory-scale batch assays	[195]

Chen et al. [182] raised a question worthy for future researches related to the possible improvement of the removal efficiency of pollutants by algae under optimal light conditions and after their acclimation.

The final step of the treatment, harvesting, which represents a significant amount of the production cost and energy consumption, is one of the most challenging processes. One of the most environmentally friendly methods is bioflocculation [150,151]; nevertheless, it may not be efficient enough due to the small size of microalgae cells. Other technologies, namely, co-pelletization through the addition of other microorganisms (bacterial and fungal species) to microalgae culture, which do not require the addition of chemicals or inputs of energy, may be promising solutions. However, further research on this topic is recommended.

Such as for mycoremediation, genetic tools can represent an essential step to improve algae performance. However, genetically modified algae for bioremediation are rarely reported. An example is a small laccase from *Streptomyces coelicolor* that was engineered by structure-based design and site-directed mutagenesis to improve its activity on commercially relevant substrates. The variants generated showed up to a 40-fold increased efficiency on 2,6-dimethoxyphenol, as well as the ability to use mediators with considerably higher redox potentials [195].

3. Biosorption

Adsorption includes all the processes involving a physical or chemical interaction between the surface of a solid material (adsorbent) and the pollutant (adsorbate); biosorption is a subcategory of adsorption, and may be simply defined as the removal of substances from solution by biological material (live or dead). It is a property of living and dead biomass to bind and abiotically concentrate compounds [196,197]. In addition to being an efficient and low-cost process, biosorption offers advantages over conventional processes, and avoids the use of chemicals, namely, nutrient supply [197].

3.1. Biosorption Materials

Activated carbon is the most widely and effectively used sorbent, since it has a porous structure consisting of a network of interconnected macropores, mesopores, and micropores that provide a good capacity for the adsorption of organic molecules due to its high specific surface area. However, it is quite expensive, United States (US) $20–22/kg [198], and the higher the quality is, the greater the cost is [199,200], which limits its widespread use. According to the Swedish Environmental Protection Agency [201], the costs depend mainly on the size of the treatment plant. For facilities using granular activated carbon, for 2000–10,000 population equivalent, the following costs are associated: 0.72 million US $ for installation, 0.04 million US $/year of capital expenditure, 0.1 million US $/year of operating expenditure, and less than 0.01 kWh/m^3 of operational electricity consumption, corresponding to a total cost of 0.09–0.11 US $/m^3 of treated wastewater. It is verified that the annual operating expenditure has a major contribution to the total cost and that the major fraction of the operating costs is due to activated carbon, which demonstrates the need to look for low-cost adsorbents. This led to the search for low-cost sorbents, which require little processing and that are abundant in nature [202]. In this search, biosorbents from microorganisms were shown to be a promising alternative since, due to their reduced size, they present high specific surface area and they were shown to be able to sorb different organic and inorganic pollutants from solutions [203]. A wide range of microorganisms were studied in biosorption processes; these include microalgae and fungi. The complexity of the microorganism structure implies that there are many ways for the contaminants to be captured by the cells; in some cases, these are still not very well understood. The cell wall is the first component that meets the pollutants, where the solutes can be deposited on the surface or within the cell-wall structure.

Fungal cell walls are structurally complex with several functional groups such as carboxyl, hydroxyl, amino, sulfonate, and phosphonate, which bring about the excellent adsorption properties of fungi [204–208].

Algal cell walls exhibit some variations in their structure. The main groups present are amino, amine, hydroxyl, imidazole, phosphate, and sulfate [209]. This availability of functional groups minimizes the selectivity problems and allows the high efficiency of removal to be more easily achieved [210]. For example, Navarro et al. [211] proposed the formation of hydrogen bonds as the main mechanism for the removal of sulfamethoxazole and sulfacetamide by marine algae *Lessonia nigrescens* Bory and *Macrocystis integrifolia* Bory.

3.2. Treatment Systems

Biosorption can proceed in batch or continuous mode. Both modes are frequently employed to conduct laboratory-scale experiments, while continuous mode is the most employed for industrial applications. Packed-bed, fluidized-bed, and continuous-stirred-bed reactors are three types of design used in biosorption experiments. Several studies showed that packed-bed columns are the most suitable for liquid–solid separation, and scaling was found to be minimal [212,213]. Fluidized-bed and continuous-stirred-bed reactors are only occasionally used, since fluidized-bed reactors require a high flow rate, which is sometimes difficult to achieve, and stirred-bed reactors require that biomass be in powdered form; beyond that, they have high associated costs and high requirements of operation and maintenance [207,214].

In the last few decades, many patents were developed, focused on improving the sorption capacity of biosorbents through their modification or immobilization, mainly for metal removal [197]. However, further attention should be given to the removal of pharmaceuticals.

3.3. Mechanisms of Removal

Biosorption is a physico-chemical process that involves several mechanisms (e.g., adsorption, ion exchange, surface complexation, and microprecipitation), as shown in Figure 4 [215]. Biosorbent surfaces are characterized by active, energy-rich sites that are able to interact with compounds in the adjacent aqueous phase due to their specific electronic and spatial properties. Sorption occurs because, from the thermodynamic point of view, the molecules prefer to be in a low-energy state. A molecule sorbed onto a surface has a lower energetic state on a surface than in the aqueous phase. Therefore, the molecule is attracted to the surface and to a lower-energy state. The attraction of a molecule to a surface can be caused by physical and/or chemical forces. Electrostatic forces govern the interactions between most sorbates and biosorbents. These forces include dipole–dipole interactions or London/van der Waals forces, and hydrogen bonds [216].

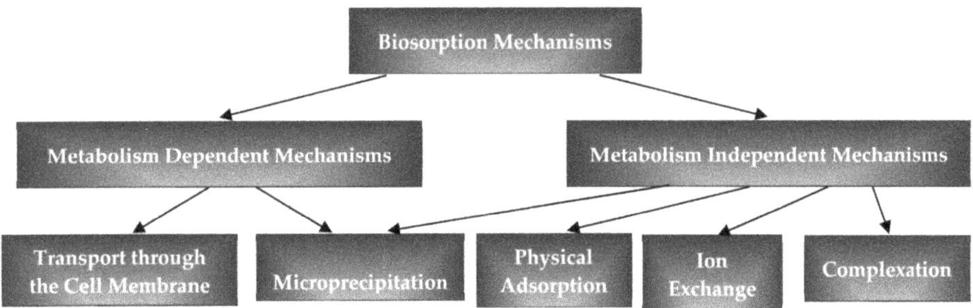

Figure 4. Biosorption mechanisms.

Biosorption equilibrium is not established instantaneously. The mass transfer from the solution to the sorption sites within the adsorbent particles is constrained by mass transfer resistances that determine the time required to reach the state of equilibrium. The rate of adsorption is usually limited by diffusion processes toward the external adsorbent surface and within the porous adsorbent particles.

The progress of the biosorption process includes four consecutive steps: transport of the sorbate from the bulk liquid phase to the hydrodynamic boundary layer localized around the biosorbent particle; transport through the boundary layer to the external surface of the biosorbent, termed film diffusion or external diffusion; pore diffusion and/or surface diffusion toward the interior of the biosorbent particle (termed intraparticle diffusion or internal diffusion); and energetic interaction between the sorbate molecules and the sorption sites.

Kinetic and equilibrium data are modeled using different approaches in order to explain the biosorption mechanism of the pollutants' removal [217]. The equilibrium distribution on the sorbed pollutant between the biosorbent and the aqueous phase is required to determine the maximum biosorbent's uptake capacity. The sorption kinetics provides additional important information about the sorption mechanism, especially the rate of pollutant removal. In wastewater treatment, kinetics information is important for setting an optimum residence time of the wastewater at the biosolid phase interface.

Typically, the dependence of the sorbed amount on the equilibrium concentration is determined experimentally at constant temperature, and the measured data are subsequently described by an appropriate isotherm equation.

For fixed-bed columns, the dynamic behavior and efficiency are described in terms of the effluent/affluent concentration as a function of time (or volume of treated liquid), i.e., the breakthrough curve. Experimental "breakthrough curve" determination allows verifying the applicability of a chosen adsorption model for a given biosorbent/sorbate system and estimating the related mass transfer coefficients [217].

3.4. Factors That Affect the Process

Several factors affect biosorption; most of them are common to adsorption processes, like pH, temperature, contaminant concentration, nature of sorbent and sorbate, contact time, ionic strength, the presence of other compounds, the use of dead or alive biomass [196,214,218], and the presence/absence of metabolic processes, in the case of living cells [215]. In living organisms, metabolic processes may affect physico-chemical biosorption mechanisms, pollutant bioavailability, chemical speciation, and accumulation/transformation [196].

Organic pollutants have a wide diversity of chemical structures; therefore, factors such as the molecular size, charge, solubility, hydrophobicity, and reactivity are important factors in biosorption. Human pharmaceutical metabolites, for example, are usually more polar and hydrophilic than the parent compounds; therefore, it is expected that they will not be significantly removed by sorption [219].

By changing the properties of the liquid phase (e.g., concentration, temperature, pH), sorbed species may be released from the surface and transferred back into the liquid phase. This reverse process is referred to as desorption.

In the study of Aksu and Tunc [220], the removal of penicillin G potassium salt by dried fungus *Rhizopus arrhizus* was strongly dependent on the pH and temperature. Maximum sorption was observed at initial pH values of 6.0 and at 35 °C. They also [220] explored the possibility of using the dried fungus for the batch removal of penicillin G potassium salt from aqueous solution and compared the experimental uptake with the results obtained with powdered activated carbon, in the same conditions. The uptake capacity determined was 459.0 $mg{\cdot}g^{-1}$ and 375.0 $mg{\cdot}g^{-1}$ for dried *Rhizopus arrhizus* and for an activated carbon (commercial powdered activated carbon, SIGMA, C-4386, washed with hydrochloric acid), respectively.

The use of dead biomass has advantages compared to the use of live biomass, such as the absence of toxicity limitations and the absence of nutrient requirement for microorganism's growth; however, live biomass has the advantage of using metabolic processes that complement the overall removal process [220].

Tam et al. [188] investigated tributyltin removal from artificial wastewater by dead and live cells of four microalgal species, *Chlorella miniata, Chlorella sorokiniana, Scenedesmus dimorphus*, and *Scenedesmus platydiscus*. In general, dead cells were more efficient than live cells during the three days of exposure; however, at the end of 14 days, removal efficiencies were identical to those that were reached by live cells. More than 90% and 85% of tributyltin was removed by dead cells of *Scenedesmus* and *Chlorella*, respectively, in the first three days. After three days of treatment, total amount of sorbed tributyltin per unit of biomass for dead cells of *Chlorella miniata*, *Chlorella sorokiniana*, *Scenedesmus dimorphus*, and *Scenedesmus platydiscus* was, respectively, 11.3 ± 0.6, 4.0 ± 0.3, 1.31 ± 0.09, and 0.35 ± 0.04 $mg{\cdot}g^{-1}$, while, for live cells, it was, respectively, 0.24 ± 0.03, 0.58 ± 0.12, 0.083 ± 0.003, and 0.0603 ± 0.0005 $mg{\cdot}g^{-1}$. Many other studies allowed observing that the fast concentration decrease in the beginning of the assays was caused by the sorption process, followed by a much slower concentration decrease caused by other biological removal mechanisms. This fact is probably due to the large surface area of the individual cells, which provides a large interface between the aqueous and cell phases. Nevertheless, unlike the other removal mechanisms, sorption is an equilibrium process that ends when equilibrium is reached, which may lead to an incomplete removal of pollutants [164,180,182,189]. Combining biosorption with other removal mechanisms seems to be a successful strategy for the removal of pharmaceuticals.

3.5. Biosorption Potential as a Wastewater Treatment Technology

The potential of biosorption as a wastewater treatment technology was frequently cited in the literature, as shown in Table 2. Usually, the performance of commercial adsorbents is very good due to the high specific surface area and the high porosity, but they present high costs which prevents their wide use, especially when high flowrates are involved. The use of biosorbents, which are usually easily available, presents advantages such as their low cost, and a more eco-friendly and sustainable nature. Like bioremediation, several aspects make biosorption a promising choice, since it is a low-cost treatment, has low operating costs, does not produce chemical sludge to handle, and is highly efficient [221]. Some disadvantages associated with the process are the low specificity and low robustness of the systems [222], the impossibility of long-term use of the suspended biomass, and the difficult separation of the biomass and the treated effluent [223]. The use of packed-bed columns is considered the most suitable biosorption, allowing an easy liquid–solid separation. The use of biosorbents in granulated form instead of powdered form presents the advantage of avoiding column clogging. Furthermore, the immobilization of fungi and algae can increase the efficiency of the process as it allows their use for several cycles [224].

Table 2. Biosorption studies with fungi and algae for removal of pharmaceuticals.

Compounds	Compounds Source	Strains	Removal Mechanisms	Technologies	Reference
Ibuprofen, clofibric acid, and carbamazepine	Synthetic media	Fungi *Trametes versicolor* *Irpex lacteus* *Ganoderma lucidum* *Phanerochaete chrysosporium*	Biodegradation Adsorption	Laboratory-scale batch assays	[46]
Carbamazepine	Synthetic media	Fungi *Trametes versicolor*	Biodegradation Adsorption	Laboratory-scale batch assays Pilot-scale glass air-pulsed fluidized bioreactor (continuous and batch feed)	[47]
Carbamazepine	Synthetic media	Fungi *Phanerochaete chrysosporium*	Biodegradation Biosorption	Pilot-scale plate bioreactor (continuous and batch feed)	[49]
Carbamazepine, ibuprofen, clofibric acid, ketoprofen, metronidazole, triclosan, 17-α-ethinylestradiol, 17-β-estradiol-17-acetate, estrone, estriol, 17-β-estradiol, gemfibrozil, amitriptyline, primidone, salicylic acid, diclofenac, naproxen	Synthetic media	Fungi *Trametes versicolor*	Biodegradation Biosorption	Laboratory-scale batch assays Pilot-scale fungus-augmented membrane bioreactor (continuous feed)	[51]
Metronidazole, salicylic acid, primidone, amitriptyline, carbamazepine, ketoprofen, naproxen, ibuprofen, gemfibrozil, diclofenac, triclosan, estriol, estrone, 17-α-ethinylestradiol, 17-β-estradiol, 17-β-estradiol-17-acetate	Synthetic media	Fungi *Trametes versicolor*	Biodegradation Biosorption	Laboratory-scale batch assays	[54]
Estrone, 17β-estradiol, 17α-ethinylestradiol, and estriol	WWTP wastewater	Fungi *Trametes versicolor* *Myceliophthora thermophila*	Biodegradation Adsorption	Laboratory-scale batch assays	[57]
Sulfapyridine, sulfapyridine, and sulfamethazine	Synthetic media	Fungi *Trametes versicolor*	Biodegradation and biosorption	Laboratory-scale batch assays Pilot-scale air-pulsed fluidized bed bioreactor (continuous feed)	[59]
Diclofenac, ibuprofen, naproxen, carbamazepine, and diazepam	Synthetic media	Fungi *Phanerochaete chrysosporium*	Biodegradation Biosorption	Pilot-scale continuous stirred tank reactor and fixed-bed reactor	[64]
Diclofenac, ketoprofen, and atenolol	Hospital wastewater	Fungi *Pleurotus ostreatus*	Biodegradation Adsorption	Pilot-scale air-pulsed fluidized-bed bioreactor (continuous and batch feed) Laboratory-scale batch assays	[67]

Table 2. *Cont.*

Compounds	Compounds Source	Strains	Removal Mechanisms	Technologies	Reference
17α-ethinylestradiol and carbamazepine	Synthetic media	Fungi *Pleurotus* sp. P1 *Pleurotus ostreatus* BS (unidentified) basidiomycete strain BNI	Biodegradation Adsorption	Laboratory-scale batch assays	[68]
Tetracycline	Synthetic media	Algae *Chlorella vulgaris*	Photodegradation Biosorption	Laboratory-scale batch assays Pilot-scale high-rate algal ponds (batch feed)	[164]
Progesterone and norgestrel	Synthetic media	Algae *Scenedesmus obliquus* *Chlorella pyrenoidosa*	Biodegradation Biosorption	Laboratory-scale batch assays	[168]
Carbamazepine	Synthetic media	Algae *Chlamydomonas Mexicana* *Scenedesmus obliquus*	Biodegradation Adsorption Bioaccumulation	Laboratory-scale batch assays	[169]
Acetaminophen, ibuprofen, ketoprofen, naproxen, carbamazepine, diclofenac, and triclosan	WWTP wastewater	Algae *Lessonia nigrescens Bory* *Macrocystis integrifolia Bory*	Biodegradation Photodegradation Biosorption	Pilot-scale high-rate algal ponds	[175]
17α-ethinylestradiol, estrone, and 17β-estradiol	Synthetic media	Algae *Anabaena cylindrical* *Chlorococcus* *Spirulina platensis, Chlorella* *Scenedesmus quadricauda* *Anaebena* (and duckweed *Lemna*)	Biodegradation Adsorption	Laboratory-scale batch assays Pilot-scale plug flow reactor (continuous feed)	[180]
Triclosan	Synthetic media	Algae *Chlorella Pyrenoidosa*	Biodegradation Biosorption	Laboratory-scale batch assays	[182]
Tributyltin	Synthetic media	Algae *Chlorella miniata* *Chlorella sorokiniana* *Scenedesmus dimorphus* *Scenedesmus platydiscus*	Biodegradation Adsorption Absorption	Laboratory-scale batch assays	[188]
Trimethoprim, sulfamethoxazole, and triclosan	Synthetic media	Algae *Nannochloris* Sp.	Biodegradation Adsorption	Laboratory-scale batch assays	[189]

Table 2. *Cont.*

Compounds	Compounds Source	Strains	Removal Mechanisms	Technologies	Reference
Diclofenac, ibuprofen, paracetamol, metoprolol, carbamazepine and trimethoprim, estrone, 17β-estradiol, and ethinylestradiol	Synthetic media	Algae *Chlorella sorokiniana*	Biodegradation Photolysis Biosorption	Laboratory-scale batch assays	[190]
β-estradiol and 17α-ethynylestradiol	WWTP anaerobic sludge	Algae *Selenastrum capricornutum* *Chlamydomonas reinhardtii*	Biodegradation Photodegradation Adsorption	Laboratory-scale batch assays	[194]
Sulfamethoxazole and sulfacetamide	Synthetic media	Algae *Lessonia nigrescens Bory* *Macrocystis integrifolia Bory*	Adsorption	Laboratory-scale batch assays	[211]
Penicillin G	Synthetic media	Fungi *Rhizopus arrhizus*	Biosorption	Laboratory-scale batch assays	[211]
Chloramphenicol, acetyl salicylic acid, clofibric acid	Synthetic media	Algae *Pterocladia capillacea* *Ulva lactuca*	Biosorption	Laboratory-scale batch assays	[225]
Paracetamol	Synthetic media	Algae Alga *Synechocystis* sp.	Biosorption	Laboratory-scale batch assays Pilot-scale continuous bubbling column photobioreactor (continuous feed)	[226]

4. Conclusions and Future Perspectives

The urban wastewater treatment industry is facing challenges, such as the fate of pharmaceuticals, which induce the development of wastewater treatment alternatives. In this review, an overview was drawn of fungi and algae potential to remove pharmaceuticals by bioremediation and biosorption processes from aquatic matrices. Algae present interesting advantages when compared with fungi, as they have a very fast growth and are able to remove both pharmaceuticals and nutrients, which is quite important since tertiary treatments are often needed for this purpose in domestic wastewater treatment. Moreover, this biomass is a valuable bioresource for the production of biofuel or high-value by-products. Bioremediation and biosorption studies proved to be eco-friendly and low-cost promising alternatives for pharmaceutical removal when compared to conventional methods applied for the same purpose. However, most of these studies remain at a laboratory-scale and are performed with synthetic media when it is well known that wastewaters are complex matrices that differ between WWTPs and can vary per geographical region. Only a few researchers are working on real field applications, since extensive research is still needed in order to understand the complexity of the processes, their dependence on physico-chemical and biological factors, and the mechanisms involved, in order to ensure high efficiencies for pharmaceutical removal. New knowledge in genetic engineering should be introduced in order to select and amplify the most effective algae or fungi strains for pharmaceutical removal. Several microorganisms contain key metabolic genes that could be introduced into other organisms. Genetically modified fungi and algae, armed with new or increased capacities for degrading various compounds, will likely have an important future in this field since it will allow these microorganisms to effectively remove pharmaceuticals from wastewaters and also allow their continuous use, considering legal limitations, as well as a pragmatic market and cost rationale.

Author Contributions: A.S. has gathered all the information for this review paper and was helped by C.D.-M., S.A.F. and O.M.F. in the organization of the information, synthesis and critical analysis.

Funding: This research was funded by the Associate Laboratory for Green Chemistry-LAQV which is financed by national funds from FCT/MCTES (UID/QUI/50006/2019) and co-financed by the ERDF under the PT2020 Partnership Agreement (POCI-01-0145-FEDER-007265). This research was funded also by the EU and FCT/UEFISCDI/FORMAS, in the frame of the collaborative international consortium REWATER financed under the ERA-NET Cofund WaterWorks2015 Call, this ERA-NET is an integral part of the 2016 Joint Activities developed by the Water Challenges for a Changing World Joint Program Initiative (Water JPI). Andreia Silva would like to thank FCT for her Ph.D. Grant SFRH/BD/138/780/2018.

Acknowledgments: The authors are greatly indebted to all financing sources.

Conflicts of Interest: The authors declare no conflict of interest.

References

1. Arnold, K.E.; Brown, A.R.; Ankley, G.T.; Sumpter, J.P. Medicating the environment: Assessing risks of pharmaceuticals to wildlife and ecosystems. *Philos. Trans. R. Soc. B Boil. Sci.* **2014**, *369*. [CrossRef] [PubMed]
2. Oulton, R.L.; Kohn, T.; Cwiertny, D.M. Pharmaceuticals and personal care products in effluent matrices: A survey of transformation and removal during wastewater treatment and implications for wastewater management. *J. Environ. Monit.* **2010**, *12*, 1956–1978. [CrossRef] [PubMed]
3. Paíga, P.; Santos, L.H.; Amorim, C.G.; Araújo, A.N.; Montenegro, M.C.; Pena, A.; Delerue-Matos, C. Pilot monitoring study of ibuprofen in surface waters of north of Portugal. *Environ. Sci. Pollut. Res.* **2013**, *20*, 2410–2420. [CrossRef] [PubMed]
4. Paíga, P.; Santos, L.H.M.L.M.; Ramos, S.; Jorge, S.; Silva, J.G.; Delerue-Matos, C. Presence of pharmaceuticals in the Lis river (Portugal): Sources, fate and seasonal variation. *Sci. Total Environ.* **2016**, *573*, 164–177. [CrossRef] [PubMed]
5. Pomati, F.; Castiglioni, S.; Zuccato, E.; Fanelli, R.; Vigetti, D.; Rossetti, C.; Calamari, D. Effects of a complex mixture of therapeutic drugs at environmental levels on human embryonic cells. *Environ. Sci. Technol.* **2006**, *40*, 2442–2447. [CrossRef] [PubMed]

6. Boxall, A.B.A.; Rudd, M.A.; Brooks, B.W.; Caldwell, D.J.; Choi, K.; Hickmann, S.; Innes, E.; Ostapyk, K.; Staveley, J.P.; Verslycke, T.; et al. Pharmaceuticals and personal care products in the environment: What are the big questions? *Environ. Health. Perspect.* **2012**, *120*, 1221–1229. [CrossRef] [PubMed]
7. Directive 2000/60/EC of the European Parliament and of the Council of 23 October 2000 Establishing a Framework for Community Action in the Field of Water Policy. Available online: https://eur-lex.europa.eu/resource.html?uri=cellar:5c835afb-2ec6-4577-bdf8-756d3d694eeb.0004.02/DOC_1&format=PDF (accessed on 4 June 2019).
8. Decision No 2455/2001/EC of the European Parliament and of the Council of 20 November 2001 Establishing the List of Priority Substances in the Field of Water Policy, and Amending Directive 2000/60/EC. Available online: http://ec.europa.eu/environment/ecolabel/documents/prioritysubstances.pdf (accessed on 4 June 2019).
9. Directive 2008/105/EC of the European Parliament and of the Council of 16 December 2008 on Environmental Quality Standards in the Field of Water Policy, Amending and Subsequently Repealing Council Directives 82/176/EEC, 83/513/EEC, 84/156/EEC, 84/491/EEC, 86/280/EEC and Amending Directive 2000/60/EC of the European Parliament and of the Council. Available online: https://eur-lex.europa.eu/legal-content/EN/TXT/PDF/?uri=CELEX:02008L0105-20130913&from=DE (accessed on 4 June 2019).
10. Directive 2013/39/EU of the European Parliament and of the Council of 12 August 2013 Amending Directives 2000/60/EC and 2008/105/EC as Regards Priority Substances in the Field of Water Policy. Available online: http://eur-lex.europa.eu/LexUriServ/LexUriServ.do?uri=OJ:L:2013:226:0001:0017:EN:PDF (accessed on 4 June 2019).
11. Commission Implementing Decision (EU) 2015/495 of 20 March 2015 Establishing a Watch list of Substances for Union-Wide Monitoring in the Field of Water Policy Pursuant to Directive 2008/105/EC of the European Parliament and of the Council (Notified Under Document C(2015) 1756). Available online: https://eur-lex.europa.eu/legal-content/EN/TXT/PDF/?uri=CELEX:32015D0495&from=PT (accessed on 4 June 2019).
12. Commission Implementing Decision (EU) 2018/840 of 5 June 2018 Establishing a Watch List of Substances for Union-Wide Monitoring in the Field of Water Policy Pursuant to Directive 2008/105/EC of the European Parliament and of the Council and repealing Commission Implementing Decision (EU) 2015/495. Available online: https://eur-lex.europa.eu/legal-content/EN/TXT/PDF/?uri=CELEX:32018D0840&rid=7 (accessed on 4 June 2019).
13. Deziel, N. Pharmaceuticals in wastewater treatment plant effluent waters. *Sch. Horiz. Univ. Minn. Morris Undergrad. J.* **2014**, *1*, 12.
14. Moller, P.; Dulski, P.; Bau, M.; Knappe, A.; Pekdeger, A.; Sommer-von Jarmerasted, C. Anthropogenic gadolinium as a conservative tracer in hydrology. *J. Geochem. Explor.* **2000**, *69*, 409–414. [CrossRef]
15. Buerge, I.J.; Poiger, T.; Muller, M.D.; Buser, H.R. Caffeine, an anthropogenic marker for wastewater contamination of surface waters. *Environ. Sci. Technol.* **2003**, *37*, 691–700. [CrossRef]
16. Martínez-Carballo, E.; González-Barreiro, C.; Scharf, S.; Gans, O. Environmental monitoring study of selected veterinary antibiotics in animal manure and soils in Austria. *Environ. Pollut.* **2007**, *148*, 570–579. [CrossRef]
17. Kim, K.R.; Owens, G.; Kwon, S.I.; So, K.H.; Lee, D.B.; Ok, Y.S. Occurrence and Environmental Fate of Veterinary Antibiotics in the Terrestrial Environment. *Water Air Soil Pollut.* **2011**, *214*, 163–174. [CrossRef]
18. Ji, K.; Kim, S.; Han, S.; Seo, J.; Lee, S.; Park, Y.; Choi, K.; Kho, Y.L.; Kim, P.G.; Park, J. Risk Assessment of chlortetracycline, oxytetracycline, sulfamethazine, sulfathiazole and erythromycin in aquatic environment. Are the current environmental concentrations safe? *Ecotoxicology* **2012**, *21*, 2031–2050. [CrossRef] [PubMed]
19. Galichet, L.Y.; Moffat, A.C.; Osselton, M.D.; Widdop, B. *Clarke´s Analysis of Drugs and Poisons*, 3rd ed.; Pharmaceutical Press: London, UK, 2004.
20. Verlicchi, P.; Al Aukidy, M.; Zambello, E. Occurrence of pharmaceutical compounds in urban wastewater: Removal, mass load and environmental risk after a secondary treatment—A review. *Sci. Total Environ.* **2012**, *429*, 123–155. [CrossRef] [PubMed]
21. Saussereau, E.; Lacroix, C.; Guerbet, M.; Cellier, D.; Spiroux, J.; Goullé, J.P. Determination of levels of current drugs in hospital and urban wastewater. *Bull. Environ. Contam. Toxicol.* **2013**, *91*, 171–176. [CrossRef] [PubMed]
22. Petrovic, M.; Gonzalez, S.; Barceló, D. Analysis and removal of emerging contaminants in wastewater and drinking water. *Trends Anal. Chem.* **2003**, *22*, 685–696. [CrossRef]

23. Hörsing, M.; Ledin, A.; Grabic, R.; Fick, J.; Tysklind, M.; Jansen, J.L.C.; Andersen, H.R. Determination of sorption of seventy-five pharmaceuticals in sewage sludge. *Water Res.* **2011**, *45*, 4470–4482. [CrossRef] [PubMed]
24. Benotti, M.; Trenholm, R.A.; Vanderford, B.J.; Holady, H.C.; Stanford, B.D.; Snyder, S.A. Pharmaceuticals and endocrine disrupting compounds in U.S. drinking water. *Environ. Sci. Technol.* **2009**, *43*, 597–603. [CrossRef] [PubMed]
25. Joss, A.; Siegrist, H.; Ternes, T.A. Are we about to upgrade wastewater treatment for removing organic micropollutants? *Water Sci. Technol.* **2008**, *57*, 251–255. [CrossRef]
26. Wu, S.; Zhang, L.; Chen, J. Paracetamol in the Environment and its Degradation by Microorganisms. *Appl. Microbiol. Biotechnol.* **2012**, *96*, 875–884. [CrossRef]
27. Sahu, O. Reduction of organic and inorganic pollutant from waste water by algae. *Int. Lett. Nat. Sci.* **2014**, *8*, 1–8. [CrossRef]
28. Boopathy, R. Factors limiting bioremediation technologies. *Bioresour. Technol.* **2011**, *74*, 63–67. [CrossRef]
29. Kartheek, B.R.; Maheswaran, R.; Kumar, G.; Banu, G.S. Biodegradation of Pharmaceutical Wastes Using Different Microbial Strains. *Int. J. Pharm. Biol. Arch.* **2011**, *2*, 1401–1404.
30. Gillespie, I.M.M.; Philip, J.C. Bioremediation, an environmental remediation technology for the bioeconomy. *Trends Biotechnol.* **2013**, *31*, 329–332. [CrossRef] [PubMed]
31. Misal, S.A.; Lingojwar, D.P.; Shinde, R.M.; Gawai, K.R. Purification and characterization of azoreductase from alkaliphilic strain Bacillus badius. *Process Biochem.* **2011**, *46*, 1264–1269. [CrossRef]
32. Mashi, B.H. Biorremediation: Issues and Challenges. *JORIND* **2013**, *11*, 1596–8303.
33. Andrade, J.A.; Augusto, F.; Jardim, I.C.S.F. Biorremediação de solos contaminados por petróleo e seus derivados. *Jornal Eclética Química* **2010**, *35*. [CrossRef]
34. Gaylarde, C.C.; Bellinaso, M.L.; Manfilo, G.P. Biorremediação - aspetos biológicos e técnicos da biorremediação de xenobióticos. *Biotecnol. Ciência Desenvolv.* **2005**, *34*, 36–43.
35. Alexopoulos, C.J.; Mims, C.W.; Blackwell, M. *Introductory Mycology*, 4th ed.; John Wiley: New York, NY, USA, 1996.
36. Badia-Fabregat, M.; Lucas, D.; Gros, M.; Rodríguez-Mozaz, S.; Barceló, D.; Caminal, G.; Vicent, T. Identification of some factors affecting pharmaceutical active compounds (PhACs) removal in real wastewater. Case study of fungal treatment of reverse osmosis concentrate. *J. Hazard. Mater.* **2015**, *283*, 663–671. [CrossRef]
37. Zhang, Y.; Xie, J.; Liu, M.; Tian, Z.; He, Z.; van Nostrand, J.D.; Ren, L.; Zhou, J.; Yang, M. Microbial community functional structure in response to antibiotics in pharmaceutical wastewater treatment systems. *Water Res.* **2013**, *47*, 6298–6308. [CrossRef]
38. Durairaj, P.; Malla, S.; Nadarajan, S.P.; Lee, P.G.; Jung, E.; Park, H.H.; Kim, B.G.; Yun, H. Fungal cytochrome P450 monooxygenases of *Fusarium oxysporum* for the synthesis of ω-hydroxy fatty acids in engineered *Saccharomyces cerevisiae*. *Microb. Cell Fact.* **2015**, *14*, 45. [CrossRef]
39. Jebapriya, G.R.; Gnanadoss, J.J. Bioremediation of textile dye using white-rot fungi: A review. *Int. J. Curr. Res. Rev.* **2013**, *5*, 1–13.
40. Morel, M.; Meux, E.; Mathieu, Y.; Thuillier, A.; Chibani, K.; Harvengi, L.; Jacquot, J.P.; Gelhaye, E. Xenomic networks variability and adaptation traits in wood decaying fungi. *Microb. Biotechnol.* **2013**, *6*, 248–263. [CrossRef] [PubMed]
41. Anastasi, A.; Tigini, V.; Varese, G.C. The bioremediation potential of different ecophysiological groups of fungi. In *Fungi as Bioremediators*; Goltapeh, E.M., Danesh, Y.R., Varma, A., Eds.; Springer-Verlag: New Delhi, India, 2013; Volume 32, pp. 29–49.
42. Spina, F.; Anastasi, A.; Prigione, V.; Tigini, V.; Varese, G.C. Biological treatment of industrial wastewaters: A fungal approach. *Chem. Eng. Trans.* **2012**, *27*, 175–180.
43. Crognale, S.; Federici, F.; Petruccioli, M. Enhanced separation of filamentous fungi by ultrasonic field: Possible usages in repeated batch processes. *J. Biotechnol.* **2002**, *97*, 191–197. [CrossRef]
44. Chisti, Y.; Moo-Young, M. On Bioreactors. In *Encyclopedia of Physical Science and Technology*; Meyers, R.A., Ed.; Academic Press: San Diego, CA, USA, 2002; Volume 2, pp. 247–271.
45. Zhong, J.J. Bioreactor Engineering. In *Comprehensive Biotechnology*; Elsevier B.V.: Amsterdam, The Netherlands, 2011; Volume 3, pp. 653–658.

46. Marco-Urrea, E.; Perez-Trujillo, M.; Vicent, T.; Caminal, G. Ability of white-rot fungi to remove selected pharmaceuticals and identification of degradation products of ibuprofen by *Trametes versicolor*. *Chemosphere* **2009**, *74*, 765–772. [CrossRef]
47. Jelic, A.; Cruz-Morato, C.; Marco-Urrea, E.; Sarra, M.; Perez, S.; Vicent, T.; Petrovic, M.; Barcelo, D. Degradation of carbamazepine by *Trametes versicolor* in an air pulsed fluidized bed bioreactor and identification of intermediates. *Water Res.* **2012**, *46*, 955–964. [CrossRef]
48. Rodarte-Moralez, A.I.; Feijoo, G.; Moreira, M.T.; Lema, J.M. Degradation of selected pharmaceutical and personal care products (PPCPs) by white-rot fungi. *World J. Microbial Biotechnol.* **2011**, *27*, 1839–1846. [CrossRef]
49. Zhang, Y.; Geißen, S.U. Elimination of carbamazepine in a nonsterile fungal bioreactor. *Bioresour. Technol.* **2012**, *112*, 221–227. [CrossRef]
50. Cruz-Morató, C.; Ferrando-Climent, L.; Rodriguez-Mozaz, S.; Barceló, D.; Marco-Urrea, E.; Vicent, T.; Sarrà, M. Degradation of pharmaceuticals in non-sterile urban wastewater by *Trametes versicolor* in a fluidized bed bioreactor. *Water Res.* **2013**, *47*, 5200–5210. [CrossRef]
51. Nguyen, L.N.; Hai, F.I.; Yang, S.; Kang, J.; Leusch, F.D.L.; Roddick, F.; Price, W.E.; Nghiem, L.D. Removal of trace organic contaminants by an MBR comprising a mixed culture of bacteria and white-rot fungi. *Bioresour. Technol.* **2013**, *148*, 234–241. [CrossRef]
52. Cruz-Morató, C.; Lucas, D.; Llorca, M.; Rodriguez-Mozaz, S.; Gorga, M.; Petrovic, M.; Barceló, D.; Vicent, T.; Sarrà, M.; Marco-Urrea, E. Hospital wastewater treatment by fungal bioreactor: Removal efficiency for pharmaceuticals and endocrine disruptor compounds. *Sci. Total Environ.* **2014**, *493*, 365–376. [CrossRef] [PubMed]
53. Gros, M.; Cruz-Morato, C.; Marco-Urrea, E.; Longrée, P.; Singer, H.; Sarrà, M.; Hollender, J.; Vicent, T.; Rodriguez-Mozaz, S.; Barceló, D. Biodegradation of the X-ray contrast agent iopromide and the fluoroquinolone antibiotic ofloxacin by the white rot fungus *Trametes versicolor* in hospital wastewaters and identification of degradation products. *Water Res.* **2014**, *60*, 228–241. [CrossRef] [PubMed]
54. Nguyen, L.; Hai, F.i.; Yang, S.; Kang, J.; Leusch, F.; Roddick, F.; Price, W.E.; Nghiem, L.D. Removal of pharmaceuticals, steroid hormones, phytoestrogens, UV-filters, industrial chemicals and pesticides by *Trametes versicolor*: Role of biosorption and biodegradation. *Int. Biodeterior. Biodegration* **2014**, *88*, 169–175. [CrossRef]
55. Badia-Fabregat, M.; Lucas, D.; Pereira, M.A.; Alves, M.A.; Pennanen, T.; Fritze, H.; Rodríguez-Mozaz, S.; Barceló, D.; Vicent, T.; Caminal, G. Continuous fungal treatment of non-sterile veterinary hospital effluent: Pharmaceuticals removal and microbial community assessment. *Appl. Microbiol. Biotechnol.* **2016**, *100*, 2401–2415. [CrossRef] [PubMed]
56. Bodin, H.; Daneshvar, A.; Gros, M.; Hultberg, M. Effects of Biopellets Composed of Microalgae and Fungi on Pharmaceuticals Present at Environmentally Relevant Levels in Water. *Ecol. Eng. J. Ecotechnol.* **2016**, *91*, 169–172. [CrossRef]
57. Becker, D.; Rodriguez-Mozaz, S.; Insa, S.; Schoevaart, R.; Barceloó, D.; de Cazes, M.; Belleville, M.P.; Sanchez-Marcano, J.; Misovic, A.; Oehlmann, J.; et al. Removal of Endocrine Disrupting Chemicals in Wastewater by Enzymatic Treatment with Fungal Laccases. *Org. Process Res. Dev.* **2017**, *21*, 480–491. [CrossRef]
58. Blánquez, P.; Guieysse, B. Continuous biodegradation of 17β-estradiol and 17α-ethynylestradiol by *Trametes versicolor*. *J. Hazard. Mater.* **2008**, *150*, 459–462. [CrossRef] [PubMed]
59. Rodriguez-Rodriguez, C.E.; García-Galán, M.J.; Blánquez, P.; Díaz-Cruz, M.S.; Barceló, D.; Caminal, G.; Vicent, T. Continuous degradation of a mixture of sulfonamides by *Trametes versicolor* and identification of metabolites from sulfapyridine and sulfathiazole. *J. Hazard. Mater.* **2012**, *213–214*, 347–354. [CrossRef]
60. Rodriguez-Rodriguez, C.E.; Marco-Urrea, E.; Caminal, G. Degradation of Naproxen and Carbamazepine in Spiked Sludge Slurry and Solid-Phase *Trametes versicolor* Systems. *Bioresour. Technol.* **2010**, *101*, 2259–2266. [CrossRef]
61. Marco-Urrea, E.; Perez-Trujillo, M.; Cruz-Morato, C.; Caminal, G.; Vicent, T. Degradation of the Drug Sodium Diclofenac *Trametes versicolor* Pellets and Identification of Some Intermediates by NMR. *J. Hazard. Mater.* **2010**, *176*, 836–842. [CrossRef]

62. Marco-Urrea, E.; Perez-Trujillo, M.; Cruz-Morato, C.; Caminal, G.; Vicent, T. White-Rot Fungus-Mediated Degradation of the Analgesic Ketoprofen and Identification of Intermediates by HPLC-DAD-MS and NMR. *Chemosphere* **2010**, *78*, 474–481. [CrossRef] [PubMed]
63. Becker, D.; Varela Della Giustina, S.; Rodriguez-Mozaz, S.; Schoevaart, R.; Barcelo, D.; de Cazes, M.; Belleville, M.P.; Sanchez-Marcano, J.; de Gunzburg, J.; Couillerot, O.; et al. Removal of antibiotics in wastewater by enzymatic treatment with fungal laccase—Degradation of compounds does not always eliminate toxicity. *Bioresour. Technol.* **2016**, *219*, 500–509. [CrossRef] [PubMed]
64. Rodarte-Moralez, A.I.; Feijoo, G.; Moreira, M.T.; Lema, J.M. Operation of stirred tank reactors (STRs) and fixed-bed reactors (FBRs) with free and immobilized *Phanerochaete chrysosporium* for the continuous removal of pharmaceutical compounds. *Biochem. Eng. J.* **2012**, *66*, 38–45. [CrossRef]
65. Wen, X.; Jia, Y.; Li, J. Enzymatic Degradation of Tetracycline and Oxytetracycline Crude Manganese Peroxidase Prepared *Phanerochaete Chrysosporium*. *J. Hazard. Mater.* **2010**, *177*, 924–928. [CrossRef] [PubMed]
66. Watanabe, R.A.; Sales, P.T.F.; Campos, L.C.; Garcia, L.C.; Valadares, T.A.; Schimidt, M.C.; Santiago, M.F. Evaluation of the use of *Pycnoporus sanguineus* fungus for phenolics and genotoxicity decay of a pharmaceutical effluent treatment. *Ambi-Agua Taubaté* **2012**, *7*, 41–50.
67. Palli, L.; Castellet-Rovira, F.; Péerez-Trujillo, M.; Caniani, D.; Sarrá-Adroguer, M.; Gori, R. Preliminary Evaluation of *Pleurotus ostreatus* for the Removal of Selected Pharmaceuticals from Hospital Wastewater. *Biotechnol. Prog.* **2017**, *33*, 1529–1537. [CrossRef]
68. Santos, I.J.S.; Grossman, M.; Sartoratto, A.; Ponezi, A.; Durranta, L. Degradation of the Recalcitrant Pharmaceuticals Carbamazepine and 17α-Ethinylestradiol by Ligninolytic fungi. *Chem. Eng. Trans.* **2012**, *27*, 169–174.
69. Esterhuizen-Londt, M.; Schwartz, K.; Pflugmacher, S. Using aquatic fungi for pharmaceutical bioremediation: Uptake of acetaminophen by *Mucor hiemalis* does not result in an enzymatic oxidative stress response. *Fungal Biol.* **2016**, *120*, 1249–1257. [CrossRef]
70. Buchicchio, A.; Bianco, G.; Sofo, A.; Masi, S.; Caniani, D. Biodegradation of carbamazepine and clarithromycin by *Trichoderma harzianum* and *Pleurotus ostreatus* investigated by liquid chromatography–high resolution tandem mass spectrometry (FTICR MS-IRMPD). *Sci. Total Environ.* **2016**, *557*, 733–739. [CrossRef]
71. Golan-Rozen, N.; Chefetz, B.; Ben-Ari, J.; Geva, J.; Hadar, Y. Transformation of the recalcitrant pharmaceutical compound carbamazepine by *Pleurotus ostreatus*: Role of cytochrome P450 monooxygenase and manganese peroxidase. *Environ. Sci. Technol.* **2011**, *45*, 6800–6805. [CrossRef]
72. Tran, N.H.; Urase, T.; Kusakabe, O. Biodegradation Characteristics of Pharmaceutical Substances by Whole Fungal Culture *Trametes versicolor* and its Laccase. *J. Water Environ. Technol.* **2010**, *8*, 125–140. [CrossRef]
73. Pointing, S.B. Feasibility of bioremediation by white-rot fungi. *Appl. Microbiol. Biotechnol.* **2001**, *57*, 20–33. [PubMed]
74. Wesenberg, D.; Kyriakides, I.; Agathos, S.N. White-rot fungi and their enzymes for the treatment of industrial dye effluents. *Biotechnol. Adv.* **2003**, *22*, 161–187. [CrossRef] [PubMed]
75. Tortella, G.; Diez, M.; Durán, N. Fungal diversity and use in decomposition of environmental pollutants. *Crit. Rev. Microbiol.* **2005**, *31*, 197–212. [CrossRef] [PubMed]
76. Buchanan, I.D.; Nicell, J.A. Kinetics of peroxidase interactions in the presence of a protective additive. *J. Chem. Technol. Biotechnol.* **1998**, *72*, 23–32. [CrossRef]
77. Buchanan, I.D.; Han, Y.S. Assessment of the potential of *Arthromyces ramosus* peroxidase to remove phenol from industry wastewaters. *Environ. Technol.* **1999**, *21*, 545–552. [CrossRef]
78. Gasser, C.A.; Hommes, G.; Schaffer, A.; Corvini, P.F. Multi-catalysis reactions: New prospects and challenges of biotechnology to valorize lignin. *Appl. Microbiol. Biotechnol.* **2012**, *95*, 1115–1134. [CrossRef]
79. Kim, Y.J.; Nicell, J.A. Impact of reaction conditions on the laccase-catalyzed conversion of bisphenol A. *Bioresour. Technol.* **2006**, *97*, 1431–1442. [CrossRef]
80. Sayadi, S.; Ellouz, R. Roles of lignin peroxidase and manganese peroxidase from *Phanerochaete chrysosporium* in the decolorization of olive mill wastewaters. *Appl. Environ. Microbiol.* **1995**, *61*, 1098–1103.
81. Nguyen, L.N.; Hai, F.I.; Price, W.E.; Leusch, F.D.L.; Roddick, F.; McAdam, E.J.; Magram, A.F.; Long, D.; Nghiem, L.D. Continuous biotransformation of bisphenol A and diclofenac by laccase in an enzymatic membrane reactor. *Int. Biodeterior. Biodegrad.* **2014**, *95*, 25–32. [CrossRef]

82. Nguyen, L.N.; Hai, F.I.; Price, W.E.; Leusch, F.D.; Roddick, F.; Ngo, H.H.; Guo, W.; Magram, S.F.; Nghiem, L.D. The effects of mediator and granular activated carbon addition on degradation of trace organic contaminants by an enzymatic membrane reactor. *Bioresour. Technol.* **2014**, *167*, 169–177. [CrossRef] [PubMed]
83. Nguyen, L.N.; Hai, F.I.; Kang, J.; Leusch, F.D.L.; Roddick, F.; Magram, S.F.; Price, W.E.; Nghiem, L.D. Enhancement of trace organic contaminant degradation by crude enzyme extract from *Trametes versicolor* culture: Effect of mediator type and concentration. *J. Taiwan Inst. Chem. Eng.* **2014**, *45*, 1855–1862. [CrossRef]
84. Nguyen, L.N.; Hai, F.I.; Dosseto, A.; Richardson, C.; Price, W.E.; Nghiem, L.D. Continuous adsorption and biotransformation of micropollutants by granular activated carbon-bound laccase in a packed-bed enzyme reactor. *Bioresour. Technol.* **2016**, *210*, 108–116. [CrossRef] [PubMed]
85. Nguyen, L.N.; van de Merwe, J.P.; Hai, F.I.; Leusch, F.D.L.; Kang, J.; Price, W.E.; Roddick, F.; Magram, S.F.; Nghiem, L.D. Laccase-syringaldehyde-mediated degradation of trace organic contaminants in an enzymatic membrane reactor: Removal efficiency and effluent toxicity. *Bioresour. Technol.* **2016**, *200*, 477–484. [CrossRef] [PubMed]
86. Rahmani, K.; Faramarzi, M.A.; Mahvi, A.H.; Gholami, M.; Esrafili, A.; Forootanfar, H.; Farzadkia, M. Elimination and detoxification of sulfathiazole and sulfamethoxazole assisted by laccase immobilized on porous silica beads. *Int. Biodeterior. Biodegrad.* **2015**, *97*, 107–114. [CrossRef]
87. Margot, J.; Bennati-Granier, C.; Maillard, J.; Blánquez, P.; Barry, D.A.; Holliger, C. Bacterial versus fungal laccase: Potential for micropollutant degradation. *AMB Express* **2013**, *3*, 63. [CrossRef] [PubMed]
88. Ji, C.; Hou, J.; Wang, K.; Zhang, Y.; Chen, V. Biocatalytic degradation of carbamazepine with immobilized laccase-mediator membrane hybrid reactor. *J. Membr. Sci.* **2016**, *502*, 11–20. [CrossRef]
89. Ji, C.; Hou, J.; Chen, V. Cross-linked carbon nanotubes-based biocatalytic membranes for micro-pollutants degradation: Performance, stability, and regeneration. *J. Membr. Sci.* **2016**, *520*, 869–880. [CrossRef]
90. Xu, R.; Tang, R.; Zhou, Q.; Li, F.; Zhang, B. Enhancement of catalytic activity of immobilized laccase for diclofenac biodegradation by carbon nanotubes. *Chem. Eng. J.* **2015**, *262*, 88–95. [CrossRef]
91. Touahar, I.E.; Haroune, L.; Ba, S.; Bellenger, J.P.; Cabana, H. Characterization of combined cross-linked enzyme aggregates from laccase, versatile peroxidase and glucose oxidase, and their utilization for the elimination of pharmaceuticals. *Sci. Total Environ.* **2014**, *481*, 90–99. [CrossRef]
92. Nair, R.R.; Demarche, P.; Agathos, S.N. Formulation and characterization of an immobilized laccase biocatalyst and its application to eliminate organic micropollutants in wastewater. *New Biotechnol.* **2013**, *30*, 814–823. [CrossRef] [PubMed]
93. Ding, H.; Wu, Y.; Zou, B.; Lou, Q.; Zhang, W.; Zhong, J.; Lu, L.; Dai, G. Simultaneous removal and degradation characteristics of sulfonamide, tetracycline, and quinolone antibiotics by laccase-mediated oxidation coupled with soil adsorption. *J. Hazard. Mater.* **2016**, *307*, 350–358. [CrossRef] [PubMed]
94. De Cazes, M.; Belleville, M.P.; Mougel, M.; Kellner, H.; Sanchez-Marcano, J. Characterization of laccase-grafted ceramic membranes for pharmaceuticals degradation. *J. Membr. Sci.* **2015**, *476*, 384–393. [CrossRef]
95. De Cazes, M.; Belleville, M.P.; Petit, E.; Llorca, M.; Rodríguez-Mozaz, S.; de Gunzburg, J.; Barceló, D.; Sanchez-Marcano, J. Design and optimization of an enzymatic membrane reactor for tetracycline degradation. *Catal. Today* **2014**, *236*, 146–152. [CrossRef]
96. Llorca, M.; Rodríguez-Mozaz, S.; Couillerot, O.; Panigoni, K.; de Gunzburg, J.; Bayer, S.; Czaja, R.; Barceló, D. Identification of new transformation products during enzymatic treatment of tetracycline and erythromycin antibiotics at laboratory scale by an on-line turbulent flow liquid-chromatography coupled to a high resolution mass spectrometer LTQ-Orbitrap. *Chemosphere* **2015**, *119*, 90–98. [CrossRef] [PubMed]
97. Singh, R.; Sidhu, S.S.; Zhang, H.; Huang, Q. Removal of sulfadimethoxine in soil mediated by extracellular oxidoreductases. *Environ. Sci. Pollut. Res.* **2015**, *22*, 16868–16874. [CrossRef] [PubMed]
98. Singh, D.; Rawat, S.; Waseem, M.; Gupta, S.; Lynn, A.; Nitin, M.; Ramchiary, N.; Sharm, K.K. Molecular modeling and simulation studies of recombinant laccase from *Yersinia enterocolitica* suggests significant role in the biotransformation of non-steroidal anti-inflammatory drugs. *Biochem. Biophys. Res. Commun.* **2016**, *469*, 306–312. [CrossRef]
99. Li, X.; Xu, Q.M.; Cheng, J.S.; Yuan, Y.J. Improving the bioremoval of sulfamethoxazole and alleviating cytotoxicity of its biotransformation by laccase producing system under coculture of *Pycnoporus sanguineus* and *Alcaligenes Faecalis*. *Bioresour. Technol.* **2016**, *220*, 333–340. [CrossRef]

100. Chen, Y.; Stemple, B.; Kumar, M.; Wei, N. Cell surface display fungal laccase as a renewable biocatalyst for degradation of persistent micropollutants bisphenol A and sulfamethoxazole. *Environ. Sci. Technol.* **2016**, *50*, 8799–8808. [CrossRef]
101. Hofmann, U.; Schlosser, D. Biochemical and physicochemical processes contributing to the removal of endocrine-disrupting chemicals and pharmaceuticals by the aquatic ascomycete *Phoma* sp. UHH 5-1-03. *Appl. Microbiol. Biotechnol* **2016**, *100*, 2381–2399. [CrossRef]
102. Shi, L.; Ma, F.; Han, Y.; Zhang, X.; Yu, H. Removal of sulfonamide antibiotics by oriented immobilized laccase on Fe_3O_4 nanoparticles with natural mediators. *J. Hazard. Mater.* **2014**, *279*, 203–211. [CrossRef] [PubMed]
103. Margot, J.; Copin, P.J.; von Gunten, U.; Barry, D.A.; Holliger, C. Sulfamethoxazole and isoproturon degradation and detoxification by a laccase-mediator system: Influence of treatment conditions and mechanistic aspects. *Biochem. Eng. J.* **2015**, *103*, 47–59. [CrossRef]
104. Martinez, D.; Larrondo, L.F.; Putnam, N.; Gelpke, M.D.S.; Huang, K.; Chapman, J.; Helfenbein, K.G.; Ramaiya, P.; Detter, J.C.; Larimer, F.; et al. Genome sequence of lignocellulose degrading fungus *Phanerochaete chrysosporium* strain RP78. *Nat. Biotechnol.* **2004**, *22*, 695–700. [CrossRef] [PubMed]
105. Fabbrini, M.; Galli, C.; Gentili, P. Comparing the catalytic efficiency of some mediators of laccase. *J. Mol. Catal. B Enzym.* **2002**, *16*, 231–240. [CrossRef]
106. Cañas, A.I.; Camarero, S. Laccases and their natural mediators: Biotechnological tools for sustainable eco-friendly processes. *Biotechnol Adv.* **2010**, *28*, 694–705. [CrossRef]
107. Hu, M.R.; Chao, Y.P.; Zhang, G.Q.; Xue, Z.Q.; Qian, S. Laccase mediator system in the decolorization of different types of recalcitrant dyes. *J. Ind. Microbiol. Biotechnol.* **2009**, *36*, 45–51. [CrossRef] [PubMed]
108. Wong, D.W. Structure and action mechanism of ligninolytic enzymes. *Appl. Biochem. Biotechnol.* **2009**, *157*, 174–209. [CrossRef]
109. Morozova, O.V.; Shumakovich, G.P.; Shleev, S.V.; Yaropolov, Y.I. Laccase-mediator systems and their applications: A review. *Appl. Biochem. Microbiol.* **2007**, *43*, 523–535. [CrossRef]
110. Pogni, R.; Baratto, M.C.; Sinicropi, A.; Basosi, R. Spectroscopic and computational characterization of laccases and their substrate radical intermediates. *Cell Mol. Life Sci.* **2015**, *72*, 885–896. [CrossRef]
111. Weng, S.S.; Liu, S.M.; Lai, H.T. Application parameters of laccase mediator systems for treatment of sulfonamide antibiotics. *Bioresour. Technol.* **2013**, *141*, 152–159. [CrossRef]
112. Kurniawati, S.; Nicell, J.A. Efficacy of mediators for enhancing the laccase-catalyzed oxidation of aqueous phenol. *Enzyme Microb. Technol.* **2007**, *41*, 353–361. [CrossRef]
113. Fillat, U.; Prieto, A.; Camarero, S.; Martínez, Á.T.; Martínez, M.J. Biodeinking of flexographic inks by fungal laccases using synthetic and natural mediators. *Biochem. Eng. J.* **2012**, *67*, 97–103. [CrossRef]
114. Ashe, B.; Nguyen, L.N.; Hai, F.I.; Lee, D.J.; van de Merwe, J.P.; Leusch, F.D.L.; Price, W.E.; Nghiem, L.D. Impacts of redox-mediator type on trace organic contaminants degradation by laccase: Degradation efficiency, laccase stability and effluent toxicity. *Int. Biodeterior. Biodegrad.* **2016**, *113*, 169–176. [CrossRef]
115. Verma, P.; Madamwar, D. Production of lignolytic enzymes for dye decolorization by co-cultivation of white rot fungi *Pleurotus ostreatus* and *Phanerochaete chrysosporium* under solid state fermentation. *Appl. Biochem. Biotechnol* **2002**, *102–103*, 109–118. [CrossRef]
116. Zhang, Y.; Geißen, S.U. In Vitro Degradation of Carbamazepine and Diclofenac Crude Lignin Peroxidase. *J. Hazard. Mater.* **2010**, *176*, 1089–1092. [CrossRef] [PubMed]
117. Kennedy, J.F.; Cabral, J.M.S. Enzyme Immobilization. In *Biotechnology*; Rehm, H.J., Reed, G., Eds.; VCH: Weinheim, Germany, 1987; pp. 347–404.
118. Asif, M.B.; Hai, F.I.; Singh, L.; Price, W.E.; Nghiem, L.D. Degradation of Pharmaceuticals and Personal Care Products by White-Rot Fungi-a Critical Review. *Curr. Pollut. Rep.* **2017**, *3*, 88–103. [CrossRef]
119. Yang, S.; Hai, F.I.; Nghiem, L.D.; Nguyen, L.N.; Roddick, F.; Price, W.E. Removal of bisphenol A and diclofenac by a novel fungal membrane bioreactor operated under non-sterile conditions. *Int. Biodeterior. Biodegrad.* **2013**, *85*, 483–490. [CrossRef]
120. Suda, T.; Hata, T.; Kawai, S.; Okamura, H.; Nishida, T. Treatment of tetracycline antibiotics by laccase in the presence of 1-hydroxybenzotriazole. *Bioresour. Technol.* **2012**, *103*, 498–501. [CrossRef]
121. Joutey, N.T.; Bahafid, W.; Sayel, H.; Ghachtouli, N.E. Biodegradation: Involved microorganisms and genetically engineered microorganisms. In *Biodegradation—Life of Science*; Chamy, R., Rosenkranz, F., Eds.; InTechOpen: London, UK, 2013; p. 305.
122. Dujon, B. The yeast genome project: What did we learn? *Trends Genet* **1996**, *12*, 263–270. [CrossRef]

123. Wood, V.; Gwilliam, R.; Rajandream, M.A.; Lyne, M.; Lyne, R.; Stewart, A.; Sgouros, J.; Peat, N.; Hayles, J.; Baker, S.; et al. The genome sequence of *Schizosaccharomyces pombe*. *Nature* **2002**, *415*, 871–880. [CrossRef]
124. Martinez, M.; Bernal, P.; Almela, C.; Velez, D.; Garcia-Agustin, P.; Serrano, R. An engineered plant that accumulates higher levels of heavy metals than *Thlaspi caerulescens*, with yields of 100 times more biomass in mine soils. *Chemosphere* **2006**, *64*, 478–485. [CrossRef] [PubMed]
125. Nam, J.M.; Fujita, Y.; Arai, T.; Kondo, A.; Morikawa, Y.; Okada, H.; Ueda, M.; Tanaka, A. Construction of engineered yeast with the ability of binding to cellulose. *J. Mol. Catal. B Enzym.* **2002**, *17*, 197–202. [CrossRef]
126. Balcázar-López, E.; Méndez-Lorenzo, L.H.; Batista-García, R.A.; Esquivel-Naranjo, U.; Ayala, M.; Kumar, V.V.; Savary, O.; Cabana, H.; Herrera-Estrella, A.; Folch-Mallol, J.L. Xenobiotic compounds degradation by heterologous expression of a *Trametes sanguineus* laccase in *Trichoderma atroviride*. *PLoS ONE* **2016**, *11*, e0147997. [CrossRef] [PubMed]
127. Sze, P. *A Biology of the Algae*, 3rd ed.; WCB/McGraw-Hill: Boston, MA, USA, 1998.
128. Ruggiero, M.A.; Gordon, D.P.; Orrell, T.M.; Bailly, N.; Bourgoin, T.; Brusca, R.C.; Cavalier-Smith, T.; Guiry, M.D.; Kirk, P.M. Correction: A Higher Level Classification of All Living Organisms. *PLoS ONE* **2015**, *10*, e0130114. [CrossRef] [PubMed]
129. Subashchandrabose, S.R.; Ramakrishnam, B.; Megharaj, M.; Venkateswarlu, K.; Naidu, R.R. Mixotrophic Cyanobacteria and microalgae as distinctive biological agents for organic pollutant degradation. *Environ. Int.* **2013**, *51*, 59–72. [CrossRef] [PubMed]
130. Fu, W.; Chaiboonchoe, A.; Khraiwesh, B.; Nelson, D.R.; Al-Khairy, D.; Mystikou, A.; Alzahmi, A.; Salehi-Ashtiani, K. Algal Cell Factories: Approaches, Applications, and Potentials. *Mar. Drugs* **2016**, *14*, 225. [CrossRef]
131. Osundeko, O.; Dean, A.P.; Davies, H.; Pittman, J.K. Acclimation of microalgae to wastewater environments involves increased oxidative stress tolerance activity. *Plant Cell Physiol.* **2014**, *55*, 1848–1857. [CrossRef]
132. Cho, K.; Lee, C.H.; Ko, K. Use of phenol-induced oxidative stress acclimation to stimulate cell growth and biodiesel production by the oceanic microalga *Dunaliella salina*. *Algal Res.* **2016**, *17*, 61–66. [CrossRef]
133. Xiong, J.Q.; Kurade, M.B.; Jeon, B.H. Biodegradation of levofloxacin by an acclimated freshwater alga *Chlorella vulgaris*. *Chem. Eng. J.* **2017**, *313*, 1251–1257. [CrossRef]
134. Rehnstam-Holm, A.S.; Godhe, A. Genetic Engineering of Algal Species. In *Biotechnology*; Doelle, H.W., Ed.; Encyclopedia of Life Support Systems (EOLSS) Publishers: Oxford, UK, 2003.
135. Muñoz, R.; Guieysse, B. Algal–bacterial processes for the treatment of hazardous contaminants: A review. *Water Res.* **2006**, *40*, 2799–2815. [CrossRef]
136. *Office of Water, Wastewater Management Fact Sheet, Energy Conservation*; EPA 832-F-06-024; U.S. Environmental Protection Agency: Washington, DC, USA, 2006; p. 7.
137. Oswald, W.J.; Golueke, C.G. Biological transformation of solar energy. *Adv. Appl. Microbiol.* **1960**, *2*, 223–262. [CrossRef] [PubMed]
138. Polprasert, C.; Dissanayake, M.; Thanh, N. Bacterial Die-Off Kinetics in Waste Stabilization Ponds. *Water Pollut. Control Fed.* **1983**, *55*, 285–296. [CrossRef]
139. Pienkos, P.; Darzins, A. The promise and challenges of microalgal-derived biofuels. *Biofuels Bioprod. Biorefin.* **2009**, 431–440. [CrossRef]
140. Becker, E.W. Micro-algae for human and animal consumption. In *Micro-Algal Biotechnology*; Borowitzka, M.A., Borowitzka, L.J., Eds.; Cambridge University Press: Cambridge, UK, 1988; pp. 222–256.
141. Ugwu, C.U.; Aoyagi, H.; Uchiyama, H. Photobioreactors for mass cultivation of algae. *Bioresour. Technol.* **2008**, *99*, 4021–4028. [CrossRef] [PubMed]
142. Chisti, Y. Biodiesel from microalgae. *Biotechnol. Adv.* **2007**, *25*, 294–306. [CrossRef] [PubMed]
143. Garcia, J.; Green, B.F.; Lundquist, T.; Mujeriego, R.; Hernández-Mariné, M.; Oswald, W.J. Long term diurnal variations in contaminant removal in high rate ponds treating urban wastewater. *Bioresour. Technol.* **2006**, *97*, 1709–1715. [CrossRef] [PubMed]
144. Harun, R.; Singh, M.; Forde, G.M.; Danquah, M.K. Bioprocess engineering of microalgae to produce a variety of consumer products. *Renew. Sustain. Energy Rev.* **2010**, *14*, 1037–1047. [CrossRef]
145. Picot, B.; Moersidik, S.; Casellas, C.; Bontoux, J. Using diurnal variations in a high rate algal pond for management pattern. *Water Sci. Technol.* **1993**, *28*, 169–175. [CrossRef]
146. Singh, S.P.; Singh, P. Effect of temperature and light on the growth of algae species: A review. *Renew. Sustain. Energy Rev.* **2015**, *50*, 431–444. [CrossRef]

147. Chen, C.Y.; Yeh, K.L.; Aisyah, R.; Lee, D.J.; Chang, J.S. Cultivation, photobioreactor design and harvesting of microalgae for biodiesel production: A critical review. *Bioresour. Technol.* **2011**, *102*, 71–81. [CrossRef]
148. Eriksen, N.T. The technology of microalgal culturing. *Biotechnol. Lett.* **2008**, *30*, 1525–1536. [CrossRef] [PubMed]
149. Wiley, P.E.; Brenneman, K.J.; Jacobson, A.E. Improved Algal Harvesting Using Suspended Air Flotation. *Water Environ. Res.* **2009**, *81*, 702–708. [CrossRef] [PubMed]
150. Akhtar, N.; Iqbal, J.; Iqbal, M. Enhancement of lead (II) biosorption by microalgal biomass immobilized onto loofa (*Luffa cylindrica*) sponge. *Eng. Life Sci.* **2004**, *4*, 171–178. [CrossRef]
151. Mallick, N. Biotechnological potential of immobilized algae for wastewater N, P and metal removal: A review. *Biometals* **2002**, *15*, 377–390. [CrossRef] [PubMed]
152. Amaro, H.M.; Guedes, A.C.; Malcata, F.X. Advances and perspectives in using microalgae to produce biodiesel. *Appl. Energy* **2011**, *88*, 3402–3410. [CrossRef]
153. Schenk, P.M.; Thomas-Hall, S.R.; Stephens, E.; Marx, U.C.; Mussgnug, J.H.; Posten, C.; Kruse, O.; Hankamer, B. Second generation biofuels: High-efficiency microalgae for biodiesel production. *Bioenergy Res.* **2008**, *1*, 20–43. [CrossRef]
154. Taylor, R.L.; Rand, J.D.; Caldwell, G.S. Treatment with algae extracts promotes flocculation, and enhances growth and neutral lipid content in *Nannochloropsis oculata*—A candidate for biofuel production. *Mar. Biotechnol.* **2012**, *6*, 774–781. [CrossRef]
155. Van Den Hende, S.; Vervaeren, H.; Saveyn, H.; Maes, G.; Boon, N. Microalgal bacterial floc properties are improved by a balanced inorganic/organic carbon ratio. *Biotechnol. Bioeng.* **2011**, *108*, 549–558. [CrossRef]
156. Su, Y.; Mennerich, A.; Urban, B. Municipal wastewater treatment and biomass accumulation with a wastewater-born and settleable algal-bacterial culture. *Water Res.* **2011**, *45*, 3351–3358. [CrossRef]
157. Zhou, W.; Cheng, Y.; Li, Y.; Wan, Y.; Liu, Y.; Lin, X.; Ruan, R. Novel fungal palletization-assisted technology for algae harvesting and wastewater treatment. *Appl. Biochem. Biotechnol.* **2012**, *167*, 214–228. [CrossRef]
158. Zhang, J.G.; Hu, B.A. novel method to harvest microalgae via co-culture of filamentous fungi to form cell pellets. *Bioresour. Technol.* **2012**, *114*, 529–535. [CrossRef] [PubMed]
159. Gultom, S.O.; Hu, B. Review of microalgae harvesting via co-pelletization with filamentous fungus. *Energies* **2013**, *6*, 5921–5939. [CrossRef]
160. Wittenberg, K.M. Preservation of high-moisture hay in storage through the use of forage additives. *Can. J. Anim. Sci.* **1991**, *71*, 429–437. [CrossRef]
161. Milledge, J.J.; Heaven, S.A. review of the harvesting of micro-algae for biofuel production. *Rev. Environ. Sci. Biotechnol.* **2013**, *12*, 165–178. [CrossRef]
162. Olguin, E.J. Dual purpose microalgae-bacteria-based systems that treat wastewater and produce biodiesel and chemical products within a biorefinery. *Biotechnol. Adv.* **2012**, *30*, 1031–1046. [CrossRef] [PubMed]
163. Matamoros, V.; Uggetti, E.; García, J.; Bayona, J.M. Assessment of the mechanisms involved in the removal of emerging contaminants by microalgae from wastewater: A laboratory scale study. *J. Hazard. Mater.* **2015**, *301*, 197–205. [CrossRef] [PubMed]
164. De Godos, I.; Muñoz, R.; Guieysse, B. Tetracycline removal during wastewater treatment in high-rate algal ponds. *J. Hazard. Mater* **2012**, *229–230*, 446–449. [CrossRef]
165. Harms, H.; Schlosser, D.; Wick, L.Y. Untapped potential: Exploiting fungi in bioremediation of hazardous chemicals. *Nat. Rev. Microbiol.* **2011**, *9*, 177–192. [CrossRef]
166. Joss, A.; Zabczynski, S.; Gobel, A.; Hoffmann, B.; Loffler, D.; McArdell, C.S.; Ternes, T.A.; Thomsen, A.; Siegrist, H. Biological degradation of pharmaceuticals in municipal wastewater treatment: Proposing a classification scheme. *Water Res.* **2006**, *40*, 1686–1696. [CrossRef]
167. Kumar, M.S.; Kabra, A.N.; Min, B.; El-Dalatony, M.M.; Xiong, J.; Thajuddin, N.; Lee, D.S.; Jeon, B.H. Insecticides induced biochemical changes in freshwater microalga *Chlamydomonas mexicana*. *Environ. Sci. Pollut. Res.* **2016**, *23*, 1091–1099. [CrossRef]
168. Peng, F.Q.; Ying, G.G.; Yang, B.; Liu, S.; Lai, H.J.; Liu, Y.S.; Chen, Z.F.; Zhou, G.J. Biotransformation of progesterone and norgestrel by two freshwater microalgae (*Scenedesmus obliquus* and *Chlorella pyrenoidosa*): Transformation kinetics and products identification. *Chemosphere* **2014**, *95*, 581–588. [CrossRef] [PubMed]
169. Xiong, J.Q.; Kurade, M.B.; Abou-Shanab, R.A.I.; Ji, M.K.; Choi, J.J.O.; Jeon, B.H. Biodegradation of carbamazepine using freshwater microalgae *Chlamydomonas mexicana* and *Scenedesmus obliquus* and the determination of its metabolic fate. *Bioresour. Technol.* **2016**, *205*, 183–190. [CrossRef] [PubMed]

170. Maes, H.M.; Maletz, S.X.; Ratte, H.T.; Hollender, J.; Schaeffer, A. Uptake, elimination, and biotransformation of 17α-ethinylestradiol by the freshwater alga *Desmodesmus subspicatus*. *Environ. Sci. Technol.* **2014**, *48*, 12354–12361. [CrossRef] [PubMed]
171. Ding, T.; Yang, M.; Zhang, J.; Yang, B.; Lin, K.; Li, J.; Gan, J. Toxicity, degradation and metabolic fate of ibuprofen on freshwater diatom *Navicula* sp. *J. Hazard. Mater.* **2017**, *330*, 127–134. [CrossRef] [PubMed]
172. Thies, F.T.; Backhaus, T.; Bossmann, B.; Grimme, L.H. Xenobiotic biotransformation in unicellular green algae. *Plant Physiol.* **1996**, *112*, 361–370. [CrossRef] [PubMed]
173. Xiong, J.Q.; Kurade, M.B.; Patil, D.V.; Jang, M.; Paeng, K.J.; Jeon, B.H. Biodegradation and metabolic fate of levofloxacin via a freshwater green alga, *Scenedesmus obliquus* in synthetic saline wastewater. *Algal Res.* **2017**, *25*, 54–61. [CrossRef]
174. Flemming, H.C.; Wingender, J. The biofilm matrix. *Nat. Rev. Microbiol.* **2010**, *8*, 632–633. [CrossRef] [PubMed]
175. Matamoros, V.; Gutiérrex, R.; Ferrer, I.; García, J.; Bayona, J.M. Capability of microalgae-based wastewater treatment systems to remove emerging organic contaminants: A pilot-scale study. *J. Hazard. Matter.* **2015**, *288*, 34–42. [CrossRef] [PubMed]
176. Díaz-Garduño, B.; Pintado-Herrera, M.G.; Biel-Maeso, M.; Rueda-Márquez, J.J.; Lara-Martín, P.A.; Perales, J.A.; Manzano, M.A.; Garrido-Pérez, C.; Martín-Díaz, M.L. Environmental Risk Assessment of Effluents as a Whole Emerging Contaminant—Efficiency of Alternative Tertiary Treatments for Wastewater Depuration. *Water Res.* **2017**, *119*, 136–149. [CrossRef] [PubMed]
177. Hom-Diaz, A.; Jaén-Gil, A.; Bello-Laserna, I.; Rodríguez-Mozaz, S.; Vicent, T.; Barceló, D.; Blánquez, P. Performance of a Microalgal Photobioreactor Treating Toilet Wastewater: Pharmaceutically Active Compound Removal and Biomass Harvesting. *Sci. Total Environ.* **2017**, *592*, 1–11. [CrossRef] [PubMed]
178. Gentili, F.G.; Fick, J. Algal cultivation in urban wastewater: An efficient way to reduce pharmaceutical pollutants. *J. Appl. Phycol.* **2017**, *29*, 255–262. [CrossRef] [PubMed]
179. Lai, K.M.; Scrimshaw, M.D.; Lester, J.N. Biotransformation and Bioconcentration of Steroid Estrogens by *Chlorella Vulgaris*. *Appl. Environ. Microbiol.* **2002**, *68*, 859–864. [CrossRef]
180. Shi, W.; Wang, L.; Rousseau, D.P.L.; Lens, P.N.L. Removal of Estrone, 17α-Ethinylestradiol, and 17β-Estradiol in Algae and Duckweed-Based Wastewater Treatment Systems. *Environ. Sci. Pollut. Res.* **2010**, *17*, 824–833. [CrossRef] [PubMed]
181. Mani, D.; Kumar, C. Biotechnological advances in bioremediation of heavy metals contaminated ecosystems: An overview with special reference to phytoremediation. *Int. J. Environ. Sci. Technol.* **2014**, *11*, 843–872. [CrossRef]
182. Wang, S.; Wang, X.; Poon, K.; Wang, Y.; Li, S.; Liu, H.; Lin, S.; Cai, Z. Removal and Reductive Dechlorination of Triclosan by *Chlorella Pyrenoidosa*. *Chemosphere* **2013**, *92*, 1498–1505. [CrossRef]
183. Chen, J.; Zheng, F.; Guo, R. Algal feedback and removal efficiency in a sequencing batch reactor algae process (SBAR) to treat the antibiotic cefradine. *PLoS ONE* **2015**, *10*, e0133273. [CrossRef] [PubMed]
184. Greca, M.D.; Pinto, G.; Pistillo, P.; Pollio, A.; Previtera, L.; Temussi, F. Biotransformation of ethinylestradiol by microalgae. *Chemosphere* **2008**, *70*, 2047–2053. [CrossRef]
185. Zhou, G.J.; Ying, G.G.; Liu, S.; Zhou, L.J.; Chen, Z.F.; Peng, F.Q. Simultaneous removal of inorganic and organic compounds in wastewater by freshwater green microalgae. *Environ. Sci.* **2014**, *16*, 2018–2027. [CrossRef]
186. Escapa, C.; Coimbra, R.N.; Paniagua, S.; García, A.I.; Otero, M. Nutrients and pharmaceuticals removal from wastewater by culture and harvesting of *Chlorella sorokiniana*. *Bioresour. Technol.* **2015**, *185*, 276–284. [CrossRef]
187. Santos, C.E.; de Coimbra, R.N.; Bermejo, S.P.; Pérez, A.I.G.; Cabero, M.O. Comparative Assessment of Pharmaceutical Removal from Wastewater by the Microalgae *Chlorella sorokiniana*, *Chlorella vulgaris* and *Scenedesmus obliquus*. In *Biological Wastewater Treatment and Resource Recovery*; Farooq, R., Ahmad, Z., Eds.; IntechOpen: London, UK, 2017. [CrossRef]
188. Tam, N.F.Y.; Chong, A.M.Y.; Wong, Y.S. Removal of Tributyltin (TBT) by Live and Dead Microalgal Cells. *Mar. Pollut. Bull.* **2002**, *45*, 362–371. [CrossRef]
189. Bai, X.; Acharya, K. Removal of Trimethoprim, Sulfamethoxazole, and Triclosan by the Green Alga *Nannochloris* Sp. *J. Hazard. Mater.* **2016**, *315*, 70–75. [CrossRef] [PubMed]
190. De Wilt, A.; Butkovskyi, A.; Tuantet, K.; Leal, L.H.; Fernandes, T.V.; Langenhoff, A.; Zeeman, G. Micropollutant Removal in an Algal Treatment System Fed with Source Separated Wastewater Streams. *J. Hazard. Mater.* **2016**, *304*, 84–92. [CrossRef] [PubMed]

191. Dzomba, P.; Kugara, J.; Mukunyaidze, V.V.; Zaranyika, M.F. Biodegradation of tetracycline antibacterial using green algal species collected from municipal and hospital effluents. *Der. Chem. Sin.* **2015**, *6*, 27–33.
192. Escapa, C.; Coimbra, R.N.; Paniagua, S.; García, A.I.; Otero, M. Removal of pharmaceuticals from industrial wastewaters by microalgae culture. In Proceedings of the International Conference on Industrial Waste and Wastewater Treatment and Valorisation, Athens, Greece, 21–23 May 2015.
193. Xiong, J.Q.; Kurade, M.B.; Kim, J.R.; Roh, H.S.; Jeon, B.H. Ciprofloxacin toxicity and its co-metabolic removal in a freshwater microalgae *Chlamydomonas mexicana*. *J. Hazard. Mater.* **2017**, *323*, 212–219. [CrossRef] [PubMed]
194. Hom-Diaz, A.; Llorca, M.; Rodríguez-Mozaz, S.; Vicent, T.; Barceló, D.; Blánquez, P. Microalgae cultivation on wastewater digestate: β-estradiol and 17α-ethynylestradiol degradation and transformation products identification. *J. Environ. Manag.* **2015**, *155*, 106–113. [CrossRef]
195. Toscano, M.D.; De Maria, L.; Lobedanz, S.; Ostergaard, L.H. Optimization of a small laccase by active-site redesign. *ChemBioChem* **2013**, *14*, 1209–1211. [CrossRef]
196. Gadd, G.M. Biosorption: Critical review of scientific rationale, environmental importance and significance for pollution treatment. *J. Chem. Technol. Biotechnol.* **2009**, *84*, 13–28. [CrossRef]
197. Michalak, I.; Chojnacka, K.; Witek-Krowiak, A. State of the art for the biosorption process—A review. *Appl. Biochem. Biotechnol.* **2013**, *170*, 1389–1416. [CrossRef]
198. Babel, S.; Kurniawan, T.A. Low-cost adsorbents for heavy metals uptake from contaminated water: A review. *J. Hazard. Mater.* **2003**, *97*, 219–243. [CrossRef]
199. Dutta, M.; Duttam, N.N.; Bhattacharya, K.G. Aqueous phase adsorption of certain beta-lactam antibiotics onto polymeric resins and activated carbon. *Sep. Purif. Technol.* **1999**, *16*, 213–224. [CrossRef]
200. Aitcheson, S.J.; Arnett, J.; Murray, K.R.; Zhang, J. Removal of aquaculture therapeutants by carbon adsorption. 1. Equilibrium adsorption behaviour of single components. *Aquaculture* **2000**, *183*, 269–284. [CrossRef]
201. *Swedish Environmental Protection Agency, Advanced Wastewater Treatment for Separation and Removal of Pharmaceutical Residues and Other Hazardous Substances-Needs, Technologies and Impacts*; REPORT 6803; Naturvårdsverket: Stockholm, Sweden, 2017.
202. Bailey, S.E.; Olin, T.J.; Bricka, M.; Adrian, D.D. A review of potentially low-cost adsorbents for heavy metals. *Water Res.* **1999**, *33*, 2469–2479. [CrossRef]
203. Tsezos, M.; Bell, J.P. Comparison of the biosorption and desorption of hazardous organic pollutants by live and dead biomass. *Water Res.* **1989**, *23*, 561–568. [CrossRef]
204. Stumm, W.; Morgan, J.J. *Aquatic Chemistry: Chemical Equilibria and Rates in Natural Waters*, 3rd ed.; John Wiley & Sons: New York, NY, USA, 1996.
205. Gow, N.A.R.; Gadd, G.M. *The Growing Fungus*; Chapman and Hall: London, UK, 1995.
206. Gadd, G.M. Interactions of fungi with toxic metals. *New. Phytol.* **1993**, *124*, 25–60. [CrossRef]
207. Aksu, Z. Application of biosorption of the removal of organic pollutants: A review. *Process Biochem.* **2005**, *40*, 997–1026. [CrossRef]
208. Gadd, G.M.; Mowll, J.L. Copper uptake by yeast-like cells, hyphae and chlamydospores of *Aureobasidium pullulans*. *Exp. Mycol.* **1985**, *9*, 230–240. [CrossRef]
209. Crist, R.H.; Oberholser, K.; Shank, N.; Nguyen, M. Nature of bonding between metallic ions and algal cell walls. *Environ. Sci. Technol.* **1981**, *15*, 1212–1217. [CrossRef]
210. Bhatti, H.N.; Hamid, S. Removal of uranium (VI) from aqueous solutions using *Eucalyptus citriodora* distillation sludge. *Int. J. Environ. Sci. Technol.* **2014**, *11*, 813–822. [CrossRef]
211. Navarro, A.E.; Lim, H.; Chang, E.; Lee, Y.; Manrique, A.S. Uptake of Sulfa Drugs from Aqueous Solutions by Marine Algae. *Sep. Sci. Technol.* **2014**, *49*, 2175–2181. [CrossRef]
212. Chu, K.H. Improved fixed bed models for metal biosorption. *Chem. Eng. J.* **2004**, *97*, 233–239. [CrossRef]
213. Aksu, Z.; Gonen, F. Biosorption of phenol by immobilized activated sludge in a continuous packed bed: Prediction of breakthrough curves. *Process Biochem.* **2004**, *39*, 599–613. [CrossRef]
214. Vijayaraghvan, K.; Yun, Y.S. Bacterial biosorbents and biosorption. *Biotechnol. Adv.* **2008**, *26*, 266–291. [CrossRef] [PubMed]
215. Volesky, B. *Sorption and Biosorption*; BV Sorbex Inc.: Montreal-St. Lambert, QC, Canada, 2004.
216. Adamson, A.W.; Gast, A.P. *Physical Chemistry of Surfaces*, 6th ed.; Wiley: New York, NY, USA, 1997.

217. Kyzas, G.Z.; Deliyanni, E.A.; Matis, K.A.; Lazaridis, N.K.; Bikiaris, D.N.; Mitropoulos, A.C. Emerging nanocomposite biomaterials as biomedical adsorbents: An overview. *Compos. Interfaces* **2018**, *25*, 415–454. [CrossRef]
218. Park, D.; Yun, Y.S.; Park, J.M. The past, present, and future trends of biosorption. *Biotechnol. Bioproc.* **2010**, *E15*, 86–102. [CrossRef]
219. Ikehata, K.; Naghashkar, N.J.; El-Din, M.G. Degradation of aqueous pharmaceuticals by ozonation and advanced oxidation processes: A review. *Ozone Sci. Eng.* **2006**, *28*, 353–414. [CrossRef]
220. Aksu, Z.; Tunc, O. Application of biosorption for penicillin G removal: Comparison with activated carbon. *Process Biochem.* **2005**, *40*, 831–847. [CrossRef]
221. Kapoor, A.; Viraraghavan, T. Fungal biosorption—An alternative treatment option for heavy metal bearing wastewater: A review. *Bioresour. Technol.* **1995**, *53*, 195–206.
222. Eccles, H. Treatment of metal-contaminated wastes: Why select a biological process? *Trends Biotechnol.* **1999**, *17*, 462–465. [CrossRef]
223. Liu, Y.; Liu, Y.J. Biosorption isotherms, kinetics and thermodynamics. *Sep. Purif. Technol.* **2008**, *61*, 229–242. [CrossRef]
224. De Rome, L.; Gadd, G.M. Use of pelleted and immobilized yeast and fungal biomass for heavy metal and radionuclide recovery. *J. Ind. Microbiol.* **1991**, *7*, 97–104. [CrossRef]
225. El-Din, S.M.M.; Noaman, N.H.; Zaki, S.H. Removal of Some Pharmaceuticals and Endocrine Disrupting Compounds by The Marine Macroalgae. *Pterocladia capillacea* and *Ulva lactuca*. *Egypt. J. Bot.* **2017**, *57*, 139–155.
226. Escapa, C.; Coimbra, R.N.; Nuevo, C.; Vega, S.; Paniagua, S.; García, A.I.; Calvo, L.F.; Otero, M. Valorization of Microalgae Biomass by Its Use for the Removal of Paracetamol from Contaminated Water. *Water* **2017**, *9*, 312. [CrossRef]

water

Review

Advanced Oxidation Processes for the Removal of Antibiotics from Water. An Overview

Eduardo Manuel Cuerda-Correa *, **María F. Alexandre-Franco** and **Carmen Fernández-González**

Department of Organic & Inorganic Chemistry, University of Extremadura, Avda de Elvas s/n, ES-06006 Badajoz, Spain; malexandre@unex.es (M.F.A.-F.); mcfernan@unex.es (C.F.-G.)
* Correspondence: emcc@unex.es; Tel.: +34-924-489-121

Received: 31 July 2019; Accepted: 2 December 2019; Published: 27 December 2019

Abstract: In this work, the application of advanced oxidation processes (AOPs) for the removal of antibiotics from water has been reviewed. The present concern about water has been exposed, and the main problems derived from the presence of emerging pollutants have been analyzed. Photolysis processes, ozone-based AOPs including ozonation, O_3/UV, O_3/H_2O_2, and O_3/H_2O_2/UV, hydrogen peroxide-based methods (i.e., H_2O_2/UV, Fenton, Fenton-like, hetero-Fenton, and photo-Fenton), heterogeneous photocatalysis (TiO_2/UV and TiO_2/H_2O_2/UV systems), and sonochemical and electrooxidative AOPs have been reviewed. The main challenges and prospects of AOPs, as well as some recommendations for the improvement of AOPs aimed at the removal of antibiotics from wastewaters, are pointed out.

Keywords: advanced oxidation processes; antibiotics; photolysis; ozone; hydrogen peroxide; Fenton; heterogeneous photocatalysis; sonochemical oxidation; electrooxidation

---

1. Introduction

Water is a natural resource, scarce, and indispensable for human life that also allows the sustainability of the environment. It is an essential part of any ecosystem, both qualitatively and quantitatively. However, water is unevenly distributed in different regions of the world, and its quality is not the same in all of them. For example, more than one-half of the world's major rivers are severely depleted or polluted, so they degrade contaminating ecosystems and threaten the health of living beings. According to WHO and UNICEF data, 780 million people do not have access to drinking water, of which 185 million use surface water to meet their daily needs [1,2].

As noted in the second United Nations report on the development of water resources in the world [3], poor water quality slows down economic growth and can have adverse effects on health and livelihoods. Chemical contamination of surface waters, mainly due to industrial and agricultural discharges, is also a significant health risk in some developing countries. Pollution and industrial waste are endangering water resources, damaging and destroying the ecosystems of the entire world.

In recent decades, one of the biggest concerns in the environmental field is the risk associated with the pollution derived from persistent organic compounds (POPs). POPs are a group of chemical compounds that resist to a different extent the photochemical, chemical, and biochemical degradation, which causes their average life to be high in the environment. As a consequence, many POPs have been detected in low quantities ($mg{\cdot}L^{-1}$) in rivers, lakes, and oceans around the world, and even in drinking water [4]. Although the carcinogenic, mutagenic, and bactericidal properties of most POPs remain unknown, there is a great interest in their elimination from the waters to avoid their potential toxic consequences and the possible dangerous effects on the health of living organisms, including humans. Organic pollutants, not just POPs, are responsible for severe damages when they are accumulated in the environment [5].

For a long time, the scientific community has focused its efforts on the study of chemical pollutants that are regulated in different legislations. These include mostly apolar, toxic, persistent, and bioaccumulative pollutants, such as polycyclic aromatic hydrocarbons (PAHs), polychlorinated biphenyls (PCBs), or dioxins. However, in recent years, the development of new and more sensitive methods of analysis has made it possible to detect the presence of other potentially dangerous contaminants, globally referred to as "emerging pollutants". These are defined as previously unknown or unrecognized pollutants, not regulated by legislation, and whose effects on health and environment are not sufficiently known yet. They can be included in various sub-groups: Steroids and hormones, pharmaceutical and personal care products, antiseptics, surfactants, disinfection products, dyes, preservatives, etc. Their presence in the environment is not necessarily new, but concern about its possible consequences is arising since its impact on the different environments is unknown. However, due to their high production, consumption, and continuous introduction into the environment, they do not need to be persistent to cause adverse effects [6].

In the last few decades, advances in analytical methods have allowed the detection of very low concentrations (of the order of ng/L) of various compounds in waters that were not analyzed until now [7]. Even though these so-called emerging compounds are not always subjected to the existing regulations on water quality, their effects on human health and the environment make their elimination convenient [8]. The risk associated with the presence of these pollutants in the environment is not only due to their acute toxicity, but also to their genotoxicity, their capacity to develop resistance in pathogens, and the risk of endocrine alterations due to the continued exposure of aquatic organisms to these contaminants [9]. On the other hand, these products designed to be biologically active can significantly affect fishes and aquatic plants, even at very low concentrations [10]. The synergistic effect of some products on other pollutants is also known.

The degradation of the aquatic environment caused by these pollutants must be prevented [11], and control is particularly challenging due to the wide dispersion of their emission sources ranging from domestic or industrial waste to landfills. Very frequently, due to the demands of their design, they are relatively non-biodegradable compounds. For this reason, the conventional treatment of active sludge, widely used in urban wastewater treatment plants, is insufficient for the elimination of these compounds [12,13].

Therefore, it is necessary to use other technologies for the elimination of these compounds. Within these technologies, advanced oxidation processes (AOPs) have a high application potential, mainly derived from the high reactivity and low selectivity of the hydroxyl radicals. However, the presence of natural organic matter and low concentration of these micropollutants are factors to consider when applying these treatments since $\cdot OH$ radicals oxidize both substrates. Furthermore, the high reaction rate of these radicals with the micropollutants does not necessarily imply greater process efficiency [14].

Among all types of pollutants, a group of recalcitrant compounds is formed by antibiotics (ABs), which are discharged into the aquatic environment in large quantities from industrial activities or excreted by humans or animals. The accumulation of ABs in the environment constitutes a risk for the aquatic flora and fauna and may cause resistance in some bacterial strains. These compounds are tough to degrade, mainly because they tend to have a very complex structure that makes them quite stable and, consequently, poorly biodegradable [15,16]. Hence, the removal of antibiotics from aqueous medium constitutes one of the most significant challenges in the field of water treatment. A wide variety of conventional treatments have been developed to remove pollutants from waters.

These conventional treatments can be classified into three broad groups, namely physical, chemical, and biological treatments. Table 1 summarizes the advantages and disadvantages of these traditional treatments.

Table 1. Advantages and disadvantages of the different kinds of water treatments.

	Physical or Physicochemical Treatment	Biological Treatment	Chemical Treatment
Kind of pollutant	Industrial (organic, inorganic, metals)	Industrial and domestic (low concentrations of organic and some inorganic)	Industrial (organic, inorganic, metals)
Methods	Filtration Adsorption Air flotation Extraction Flocculation Sedimentation	Anaerobic Aerobic Activated muds	Thermal oxidation (combustion) Chemical oxidation Ion exchange Chemical precipitation
Advantages	Low cost of capital Relatively safe Easy to operate	Easy maintenance Relatively safe elimination of the dissolved contaminants Easy to operate	High degree of treatment Elimination of the dissolved contaminants
Disadvantages	Volatile emissions High energetic cost Complex maintenance	Volatile emissions Require elimination of residual muds Susceptible to toxins or antibiotics	High costs of capital and operation. Difficult operation

This conventional approach, hence, makes combined use of physical, chemical, and biological treatments, with the main goal of removing sediments and organic matter that could promote both, the growth of microorganisms, and the eutrophication of water bodies. Moreover, the use of conventional methods is not wholly accepted nowadays because of the high costs and operational problems [17,18]. Also, they are not very efficient for the treatment of persistent or emerging pollutants in water, such as antibiotics, since many of these compounds have complex structures and, therefore, exhibit high chemical stability that hinders their complete degradation. The generation of harmful wastes in these processes is also a significant disadvantage. Consequently, it is necessary to adopt more modern systems such as advanced oxidation processes (AOPs).

The implementation of cleaner production programs framed in the reduction of discharges and polluting effluents, and especially the application of environmentally sustainable technologies in industrial processes, is of the utmost importance nowadays. That is why the use of AOPs technologies is currently under development. AOPs were defined by Glaze et al. [19] as water treatment processes performed at pressure and temperature close to environmental conditions, which involve the generation of hydroxyl radicals in sufficient quantity to interact with the organic compounds of the medium. Wastewater treatment plants (WWTPs) generally do not reach the complete elimination of many contaminants. Therefore, they behave as an important source of release of some polluting products into the environment. The implementation of sustainable technologies is imposed as a possible solution for the recovery of high-quality treated effluent. AOPs are new water purification technologies that have been widely used in the last years due to their versatility and a broad spectrum of applicability [20] and constitute a group of very efficient methods for water and wastewater treatment [21–23].

AOPs include all the catalytic and non-catalytic processes that take advantage of the high oxidizing capacity of the hydroxyl radical (OH), and they differ from each other in the way in which this radical is generated. These processes are mainly based on the "in situ" generation of the hydroxyl radical that reacts rapidly with most organic compounds, except chlorinated alkanes [24]. Thus, such radical is generated in sufficient quantity to interact with organic compounds [4,25]. AOPs can be classified into two broad groups: Homogeneous processes and heterogeneous processes, distinguishing between those that operate with an external input of energy (radiant energy, ultrasonic energy, electrical energy) and those that do not.

Hydroxyl radicals are optimal within the group of powerful oxidants because they meet a series of requirements:

- They do not generate additional waste.
- They are not toxic and have a very short lifetime
- They are not corrosive to pieces of equipment.

- They are usually produced by assemblies that are simple to manipulate.

According to these considerations, AOPs are technologies compatible with the environment and based on them, competitive processes from an economic point of view are being developed. The viability of the AOPs depends on the efficacy of the OH radical, which is the second known species with higher oxidant power after fluorine [4,26] (Table 2).

Table 2. Standard reduction potentials in aqueous medium of the most commonly used oxidizing agents.

Oxidizer	Reduction Reaction	E°/V
Fluorine	$F_2(g) + 2H^+ + 2e^- \rightarrow 2HF$ $F_2(g) + 2e^- \rightarrow 2F^-$	3.05
Hydroxyl radical	$OH + H^+ + e^- \rightarrow H_2O$	2.80
Sulfate radical anion	$SO_4^- + e^- \rightarrow SO_4^{2-}$	2.60
Ferrate	$FeO_4^{2-} + 8H^+ + 3e^- \rightarrow Fe^{3+} + 4H_2O$	2.20
Ozone	$O_3(g) + 2H^+ + 2e^- \rightarrow O_2(g) + H_2O$	2.08
Peroxodisulfate	$S_2O_8^{2-} + 2e^- \rightarrow 2SO_4^{2-}$	2.01
Hydrogen peroxide	$H_2O_2 + 2H^+ + 2e^- \rightarrow 2H_2O$	1.76
Permanganate (a)	$MnO_4^- + 4H^+ + 3e^- \rightarrow MnO_2(s) + 2H_2O$	1.67
Hydroperoxyl radical (a)	$HO_2 + 3H^+ + 3e^- \rightarrow 2H_2O$	1.65
Permanganate (b)	$MnO_4^- + 8H^+ + 5e^- \rightarrow Mn^{2+} + 4H_2O$	1.51
Hydroperoxiyl radical (b)	$HO_2 + H^+ + e^- \rightarrow H_2O_2$	1.44
Dichromate	$Cr2O7^{2-} + 14H^+ + 6e^- \rightarrow 2Cr^{3+} + 7H_2O$	1.36
Chlorine	$Cl_2(g) + 2e^- \rightarrow 2Cl^-$	1.36
Manganese dioxide	$MnO_2 + 4H^+ + 2e^- \rightarrow Mn^{2+} + 2H_2O$	1.23
Oxygen	$O_2(g) + 4H^+ + 4e^- \rightarrow 2H_2O$	1.23
Bromine	$Br_2(l) + 2e^- \rightarrow 2Br^-$	1.07

(a) Circumneutral or weakly acidic medium; (b) Strongly acidic medium.

The ·OH radical acts in a non-selective manner on organic and organometallic contaminants in the aqueous medium, ideally leading to their complete mineralization to CO_2, water, and inorganic ions [27–31].

The ·OH radical is a highly reactive species, and therefore does not accumulate in the medium but can react efficiently with organic pollutants that are refractory to the action of other oxidants, giving rise to rate constants in the order of 10^6–10^{10} M^{-1} s^{-1} [32,33]. The most important advantage of the advanced oxidation processes is that they are respectful of the environment [34].

Hydroxyl radicals can degrade organic or organometallic compounds by three degradation mechanisms, depending on the nature of the compound:

1. Dehydrogenation or abstraction of a hydrogen atom to form water (if the substrate has C-H bonds, e.g., alkanes) and radical R· that in the presence of molecular oxygen can generate the peroxyl radical ROO· and thus initiate an oxidative sequence that can lead to mineralization, Reactions (1) and (2).

$$RH + OH \rightarrow H_2O + R \quad \text{(Reaction 1)}$$

$$R + O_2 \rightarrow ROO \quad \text{(Reaction 2)}$$

2. Hydroxylation of the organic compound by the attack of·OH in the high electron density sites, adding to the unsaturated bonds of aromatic or aliphatic compounds and initiating a chain of oxidation reactions, Reactions (3)–(6).

$$ROO\cdot + n(\cdot OH/O_2) \rightarrow xCO_2 + yH_2O \quad \text{(Reaction 3)}$$

$$ArH + OH \rightarrow ArH(OH) \quad \text{(Reaction 4)}$$

$$ArH(OH) + O_2 \rightarrow [ArH(OH)OO] \quad \text{(Reaction 5)}$$

$$[ArH(OH)OO] \rightarrow ArH(OH) + HO_2 \quad \text{(Reaction 6)}$$

3. Charge transfer by oxidation-reduction, causing the ionization of the molecule, Reaction (7).

$$OH + RX \rightarrow RX\cdot + OH^- \quad \text{(Reaction 7)}$$

The radical mechanisms are complicated so that the oxidation of organic matter by OH radicals involves several types of species and reactions:

1. Initiation reactions during which radical species R· are formed, Reaction (1).

$$RH + OH \rightarrow H_2O + R \quad \text{(Reaction 1)}$$

2. Propagation reactions involving radical species R· that react with other neutral organic molecules, Reaction (8), or with dissolved oxygen in the solution, Reaction (2).

$$R\cdot + R'H \rightarrow RH + R'\cdot \quad \text{(Reaction 8)}$$

$$R + O_2 \rightarrow ROO \quad \text{(Reaction 2)}$$

3. Termination reactions where the radicals combine, Reactions (9)–(11).

$$R\cdot + R\cdot \rightarrow R\text{-}R \quad \text{(Reaction 9)}$$

$$R\cdot + OH \rightarrow R\text{-}OH \quad \text{(Reaction 10)}$$

$$OH + OH \rightarrow H_2O_2 \quad \text{(Reaction 11)}$$

Many AOPs are based on the combination of a strong oxidizing agent (e.g., ozone or hydrogen peroxide) with a catalyst (e.g., transition metal ions or photocatalysts) and/or radiation (e.g., UV or ultrasound).

One of the possible classifications of the AOPs is based on the source of generation of the oxidizing species; that is, the method to generate the hydroxyl radicals. This classification is shown in Table 3, which lists the main types of AOPs: photolytic, based on the use of ozone, based on the use of hydrogen peroxide, photocatalytic, electrochemical, or by ultrasound. The wide variety of available techniques evidences the versatility of AOPs.

In short, the most positive characteristics of the AOPs can be summarized as follows:

- Potential capacity to carry out mineralization of organic pollutants to carbon dioxide and water, and oxidation of inorganic compounds and ions such as chlorides, nitrates, etc.
- Non-selective reactivity with the vast majority of organic compounds, especially attractive to avoid the presence of potentially toxic byproducts from the primary pollutants that can be originated by other methods that do not achieve complete oxidation.

The main disadvantage of AOPs lies in their high cost due to the use of expensive reagents (for example, H_2O_2) and energy consumption (generation of O_3 or UV radiation). Therefore, the future prospects of these processes include a renewed technology combined with adequate use of kinetic-chemical models.

Table 3. A classification of advanced oxidation processes (AOPs) according to the source of ·OH radicals.

	Generic Name	Source of OH Radicals
Advanced Oxidation Processes (AOPs)	Photolysis	UV radiation
	O_3-based processes	O_3 O_3/UV O_3/H_2O_2 O_3/H_2O_2/UV
	H_2O_2-based processes	H_2O_2/UV H_2O_2/Fe^{2+} (Fenton) H_2O_2/Fe^{3+} (Fenton-like) H_2O_2/Fe^{2+}/UV (Photo-Fenton)
	Heterogeneous photocatalysis	TiO_2/UV $TiO_2/UV/H_2O_2$
	Sonochemical oxidation	Ultrasounds 20kHz–2MHz (water sonolysis)
	Electrochemical oxidation	Electricity, 2-20A (water electrolysis)

In the following sections, a bibliographic review dealing with the most important contributions to each of the AOPs summarized in Table 3 is presented. Most of the works referred to have been published in the last decade, although references to papers published in the first decade of the 21st Century have been included, too. On the contrary, references to works published in the 1970s, 1980s, and 1990s have been reduced to a minimum. This paper follows the path of some previously published reviews dealing with the removal of antibiotics in water by AOPs [35–47].

In this review, after some preliminary considerations concerning the scarcity and necessity of water worldwide, the concerns on water pollution by persistent pollutants, and the main drawbacks of the use of conventional methods to remove them, an update is presented on the novel approaches for wastewater remediation based on advanced oxidation processes. The main challenges and future prospects of AOPs, as well as some recommendations for the improvement of AOPs aimed at the removal of antibiotics from wastewaters, have been dealt with in the final sections of this work.

2. Photolysis

Photolytic methods for the degradation of pollutants dissolved in water are based on providing energy to chemical compounds in the form of radiation, which is absorbed by different molecules to reach excited states for the time necessary to undergo different chemical reactions. Molecules absorb radiant energy in the form of quantized units called photons, which provides the energy required to excite specific electrons and form free radicals that undergo a series of chain reactions to give the reaction products. These free radicals can be generated by homolysis of weak bonds, or by electronic transfer from the excited state of the organic molecule to molecular oxygen, resulting in the superoxide radical ($O_2^{-}\cdot$), or other chemical reagents such as ozone or hydrogen peroxide (reactions that will be discussed later) so that hydroxyl radicals are produced. These photolytic methods use UV radiation due to the higher energy of their photons as indicated by the Planck's equation:

$$E_\lambda = hc/\lambda \qquad \text{ec.} \tag{1}$$

where E_λ is the energy of a photon associated with the wavelength (λ) of the radiation; h is the Planck's constant; and c is the speed of light.

Thus, direct photolysis involves the interaction of light with molecules to cause their dissociation into simpler fragments. For this reason, in any process in which UV radiation is used, photolysis could take place. The intensity and wavelength of the radiation or the quantum yield of the compound to be eliminated are factors that influence the performance of the process. As a source of UV radiation, mercury vapor lamps are usually used [48].

Indirect photodegradation is due to oxidation mediated by radicals that are generated when light excites some molecules, commonly known as *photosensitizers*. Dissolved organic matter—particularly humic and fulvic acids—and nitrate ions are two examples of photosensitizers usually found in aquatic environments. It is worth noting that the generation of radicals by sensitizers is a UV light-mediated process and, hence, indirect photodegradation takes place to the detriment of direct photolysis.

The data summarized in Table 4 illustrate the mechanisms and removal efficiencies of the different photolytic processes aimed at abating a wide variety of antibiotics in aqueous solution.

Table 4. Removal efficiency of antibiotics in waters by photodegradation.

Antibiotic	Mechanism	Maximum Removal Efficiency	Remarks	Reference
Amoxicillin (AMX) Ampicillin (AMP) Piperacillin (PPR) Penicillin V (PNV)	Mainly indirect photolysis	~100% AMX & AMP ~95% PNV ~90% PPR	Photodegradation by sunlight may play a role in the degradation of these antibiotics together with hydrolysis and microbial degradation	[49]
Cefalexin (CFL) Cefradine (CFR) Cefapirin (CFP) Cefazolin (CFZ) Cefotaxime (CFT)	Direct photolysis (CFP, CFZ) Indirect photolysis (CFL, CFR) Direct and indirect photolysis equally (CFT)	86%–89% in all cases	Photo byproducts were found to be less photolabile and more toxic than precursors	[50]
Cefradine (CFR) Cefuroxime (CFX) Ceftriaxone (CFN) Cefepime (CFM)	Direct photolysis (CFN); Indirect Photolysis (CFR, CFX, CFM)	~90% CFM ~80% CFX ~70% CFN ~60% CFR	Abiotic hydrolysis was responsible for the elimination of the cephalosporins. Direct photolysis significantly stimulated the abiotic degradation	[51]
Ceftiofur (CFF) Cefapirin (CFP)	Direct photolysis with some pH-dependent hydrolysis	~96% CFP ~92% CFF	Both compounds are relatively stable under neutral and acid environment, whereas base-catalyzed reactions (pH > 9) led to fast degradation	[52]
Ciprofloxacin (CPR)	Direct photolysis (Photooxidation, defluorination, and cleavage of the piperazine ring)	n.a.	Fast process, particularly at slightly basic pH	[53]
Difloxacin (DFL) Sarafloxacin (SRF)	Direct photolysis	>99% in both cases	SRF is the primary photoproduct of DFL and shows relatively higher persistence	[54]
Enofloxacine (ENF)	Direct and (some) indirect photolysis	Very close to 100%	Self-sensitized fluoroquinolone photooxidation via ·OH radicals and singlet oxygen also plausible	[55]

Table 4. *Cont.*

Antibiotic	Mechanism	Maximum Removal Efficiency	Remarks	Reference
Norfloxacin (NRF) Ofloxacin (OFX) Ciprofloxacin (CPR) Enrofloxacin (ENR) Sparfloxacin (SPR) Danofloxacin (DNF) Sulfanilamide (SND) Sulfaguanidine (SGD) Sulfadiazine (SDZ) Sulfamethoxazole (SLF) Sulfathiazole (STZ) Sulfisoxazole (SFX) Sulfamethizole (SMT) Sulfamethazine (SMZ) Sulfamethoxypyridazine (SMP)	Direct photolysis and collateral processes (e.g., hydrolysis)	>98.5% in all cases	Photo byproducts derived from desulfonation and/or denitrification, as well as hydroxylation of photo-oxidized heterocyclic rings were identified	[56]
Sulfamethoxazole (SLF)	Direct photolysis, hydroxylation, cleavage of the sulfonamide bond and fragmentation of the isoxazole ring	Very close to 100%	Fast process, particularly under acidic pH. Indirect photolysis results in a decrease in degradation rate	[57]
Tetracycline (TTR) Oxytetracycline (OXY) Chlortetracycline(CHL)	Mainly indirect photolysis	89.59% TTR 100% OXY 100% CHL	Effectiveness of the process is lower at higher initial concentrations for all three tetracyclines. Low concentrations of dissolved organic matter in these waters act as a photosensitizer. Higher toxicity of byproducts	[58]
Trimethoprim (TRM) Sulfamethoxazole (SLF)	Direct photolysis (48% for SLF, 18% for TRM) Indirect photolysis (52% for SLF, 82% for TRM)	~90% in both cases	Indirect photolysis is attributable to the production of ·OH radicals and triplet excited state organic matter	[59]

It is widely accepted that direct photolysis is a fast process. However, pollutants are removed by direct photolysis to a limited extent. Photosensitizers-mediated processes contribute to improving removal efficiencies.

Among the main advantages of this group of AOPs, it is worth noting that photolysis is a chemical-free treatment that requires relatively low maintenance and operational costs. Moreover, UV has proven its versatility and capacity to promote the cleavage of the chemical bonding of a wide variety of refractory compounds. Hence, the use of UV irradiation in AOPs aimed at the treatment of wastewater is rising in the last years.

However, the use of UV-alone treatments has some crucial limitations. For instance, the occurrence of organic molecules suitable to behave as photosensitizers may also cause an increase in the turbidity of the aqueous media, thus hindering the penetration of UV radiation in the polluted medium. This latter hinders the contribution of indirect photolysis so that the process becomes less efficient.

It should be noted, however, that ultraviolet radiation alone is not usually applied as an advanced oxidation process (AOP). The use of UV irradiation, together with other oxidants, may contribute to the degradation of the parental pollutants as well as that of potentially harmful byproducts [60,61]. Furthermore, most of the low-pressure UV lamps that are commercially available emit approximately 5% of the radiation of a wavelength close to 185 nm, which produces ozone in the reactor. Hence, UV has been widely used combined with oxidizing agents such as ozone or hydrogen peroxide to enhance the generation of hydroxyl radicals, as will be exposed in the next sections of this work.

In general, direct photolysis is less effective in the degradation of pharmaceuticals present in wastewaters and also requires more energy than, for example, ozonation [62]. Hence, in the next section, advanced oxidation processes based on the use of ozone are presented.

3. Ozone-Based AOPs

Ozone is a powerful oxidizing agent capable of reacting with a large number of organic and inorganic compounds. Its high oxidation potential (E° = 2.08 V, see Table 2) and the absence of the formation of dangerous byproducts during the process have increased the importance of this technique in water treatment during the past decades. The main drawback is the need to generate ozone from oxygen, for which an electric discharge over a stream of air or pure oxygen is used. This step consumes large amounts of energy, thus handicapping the scaling of the process.

3.1. Ozonation

The mechanism of oxidation by ozone is a complex process that takes place in two ways: Direct reaction with dissolved ozone (O_3) or indirect oxidation through the formation of radicals ($\cdot OH$). The extension of both mechanisms throughout the degradation of a compound depends on factors such as the nature of the contaminant, the dose of ozone, or the pH of the medium. Normally, under acidic conditions (pH < 4) direct ozonation prevails, Reaction (12):

$$3O_3 + OH^- + H^+ \rightarrow 2\,OH + 4O_2 \quad \text{(Reaction 12)}$$

On the contrary, at pH > 9, the indirect route is the most important one. As a rule, degradation rates in ozonation processes increases as pH does, since high pH favors ozone decomposition into free radicals as shown in Reaction (13). Other chemical reactions involved in the indirect oxidation with ozone are as follows:

$$O_3 + OH^- \rightarrow O_2 + HO_2^- \quad \text{(Reaction 13)}$$

$$O_3 + HO_2^- \rightarrow HO_2 + O_3^- \quad \text{(Reaction 14)}$$

$$HO_2 \rightarrow H^+ + O_2^- \quad \text{(Reaction 15)}$$

$$O_2^- + O_3 \rightarrow O_2 + O_3^- \quad \text{(Reaction 16)}$$

$$O_3^- + H^+ \rightarrow HO_3 \quad \text{(Reaction 17)}$$

$$HO_3 \rightarrow HO\cdot + O_2 \quad \text{(Reaction 18)}$$

Under alkaline conditions, however, a fast side-reaction must be taken into account:

$$HO + O_3 \rightarrow HO_2\cdot + O_2 \quad \text{(Reaction 19)}$$

This latter reaction results in a rapid generation of hydroperoxyl radicals (E° = 1.65V) to the detriment of the ·OH radicals (E° = 2.80 V) and leads to a decrease in the oxidation ability.

Table 5 lists some selected papers dealing with conventional ozonation treatments of a wide variety of antibiotics, indicating the aqueous matrix and the main experimental conditions.

Table 5. Removal efficiency of antibiotics in waters by ozonation.

Antibiotic	Matrix	Operation Conditions	Maximum Removal Efficiency	Remarks	Reference
Amoxicilin (AMX)	Deionised water	0.16 mM O_3 2.5 < pH < 7.2	~90%	Fast process, low mineralization degree	[63]
Amoxicilin (AMX)	Formulation washwater	57.5 mM O_3 3 < pH < 11.5	100%	Complete removal after 40 min treatment at pH = 11.5	[64]
Amoxicillin (AMX) Doxycycline (DXY) Ciprofloxacin (CPR) Sulphadiazine (SDZ)	Deionized water	0.003–1.5 mM O_3 pH = 6.8	70% AMX 92%–98% for DXY, CPR, and SDZ	Maximum removal achieved for 1.5 mM O_3	[65]
Azithromycin (AZT) Clarithromycin (CLR) Roxithromycin (RXT)	Spiked WWTP effluent	0.01–0.1 mM O_3 pH = 7	~99%	Excellent removal efficiencies above 0.042 mM O_3	[10]
Ciprofloxacin (CPR)	Deionized water	52 mM O_3 3 < pH < 10	95%	O_3 supply rather than reaction kinetics is rate limiting. Desethylene-CPR was identified as the major CPR degradation product	[66]
Ciprofloxacin (CPR) Erythromycin (ERY) Metronidazole (MTR) Trimethoprim (TRM)	Spiked STP effluent	0.145 mM O_3 pH = 7	100% CPR 94% ERY 100% MTR 94% TRM	Ozonation treatment was successfully used to improve conventional STP treatments	[67]
Clarithromycin (CLR) Erythromycin (ERY) Roxithromycin (RXT)	Spiked STP effluent	0.1–0.3 mM O_3 pH = 7.2	76% (CLR) 92% (ERY) 91% (RXT) Below limit of quantification in all cases	Hydroxylated antibiotics should not further promote the formation of antibiotic-resistant strains	[68]
Clarithromycin (CLR)	Distilled water	0.05 mM O_3 3.2 < pH < 4.4	100%	High rate of reaction. Antibiotics fully eliminated even at a low ozone dose	[69]
Flumequine (FLM)	Ultrapure water	140.6 mg O_3 L^{-1} (in gas phase) 3 < pH < 11	~100%	Hydroxylation, decarboxylation and defluorination were mainly involved in the FLM ozonation. Removal efficiency increases with increasing pH	[70]
Lincomycin (LNC)	Distilled water	0.06–0.10 mM O_3 2 < pH <9	~100%	Fast process, particularly at neutral pH	[71]
Lincomycin (LNC)	Distilled water	0.4 mM O_3 5.5 < pH < 7.5	~100%	Total removal achieved in 2 min	[72]
Ofloxacin (OFX) Trimethoprim (TRM) Norfloxacin (NRF) Ciprofloxacin (CPR)	Ultrapure water	0.09 mM O_3 pH = 7	~100%	All drugs completely removed within 10 s	[73]
Oxytetracycline (OXY)	Ultrapure water	0.23 mM O_3 3 < pH < 7	100%	Removal efficiency increases with increasing pH. Complete removal in 20 min	[74]
Roxithromycin (RXT)	Spiked lake, river and well water	0.002–0.042 mM O_3 pH = 8	>90%	Remarkable influence of water matrix on ozone stability, formation of radicals and scavenging	[75]

Table 5. *Cont.*

Antibiotic	Matrix	Operation Conditions	Maximum Removal Efficiency	Remarks	Reference
Spectinomycin (SPC)	Distilled water	0.06–0.10 mM O_3 2 < pH <9	~100%	Fast process, particularly at neutral pH	[71]
Sulfadiazine (SDZ) Sulfamethozaxole (SLF) Sulfapyridine (SLP) Sulfathiazole (STZ)	Spiked WWTP effluent	0.01–0.1 mM O_3 pH = 7	99%	Excellent removal efficiencies above 0.042 mM O_3	[10]
Sulfadiazine (SDZ) Sulfamethoxazole (SLF) Sulfathiazole (STZ) Sulfamethizole (SMT)	Deionized water	0.02–0.067 mM O_3 2 < pH < 10	~100%	Complete removal from contaminated water. Increasing the pH from 2.0 to 10.0 resulted in enhanced removal of the sulfonamides	[76]
Sulfamethoxazole (SLF)	Spiked lake, river and well water	0.002–0.042 mM O_3 pH = 8	>90%	Remarkable influence of water matrix on ozone stability and formation of radicals.	[75]
Sulfamethoxazole (SLF)	Spiked STP effluent	0.1–0.3 mM O_3 pH = 7.2	92% Below limit of quantification	Hydroxylated antibiotics should not further promote the formation of antibiotic-resistant strains	[68]
Sulfamethoxazole (SLF)	Distilled water	3.125–31.25 mM O_3 3 < pH < 11	100%	Removal efficiency increases with increasing pH. Complete removal in 60 min, 31.25 mM O_3	[77]
Sulfamethoxazole (SLF) Chlortetracycline (CHL)	Distilled water	O_3 concentration not provided pH = 4.63 (SLF) or 4.33 (CHL)	~100%	Total degradation achieved after 90 min. CHL was more quickly oxidized than SLF	[78]
Triclosan (TRC)	Ultrapure water	0.04 mM O_3 pH = 7	~100%	2,4-dichlorophenol, chlorocatechol, mono-hydroxy-TRC, and dihydroxy-TRC were the main byproducts. Increasing O_3 concentrations leads to decreased concentration of TRC and byproducts	[79]
Trimethoprim (TRM)	Spiked STP effluent	0.1–0.3 mM O_3 pH = 7.2	85%	Hydroxylated antibiotics should not further promote the formation of antibiotic resistant strains	[68]

If O_3 mediated oxidation is performed under acidic or near-neutral pH, the degradation of the pollutants mainly takes place through direct reactions between O_3 and the organic molecules. The main targets of ozone attack are -C = C- or -N = N- double bonds.

Several operational parameters strongly influence the formation of O_3 and its subsequent transformation into ·OH radicals. Among these parameters, the chemical structure and concentration of the pollutant, the quality of the effluent, pH (as indicated above), and temperature must be taken into consideration. The main advantages of ozonation are:

(i) The volume of effluent remains constant along the process and sludge is not formed,
(ii) Installations are relatively simple and require only a little space,
(iii) O_3 is generated in situ, so that no stock solutions of H_2O_2, iron salts, or other chemicals are needed on-site,
(iv) It can be applied even if the effluent fluctuates both in terms of flow rate and/or composition and
(v) O_3 remnants can be eliminated as ozone tends to decompose into oxygen.

As indicated above, the main inconvenience is the relatively high cost of equipment and maintenance, together with the high requirements of energy that must be supplied to carry on the process.

Another key point to be taken into consideration is the necessity to ensure an adequate mass transfer. It must be born in one's mind that O_3 molecules must be transferred from the gas phase to the liquid phase so that the attack on the chemical bonds of the organic molecules may occur.

Very frequently, the mass transfer may be regarded as the limiting step of the process. An inadequate mass transfer can negatively affect the removal efficiency of the process and, hence, result in increasing the operating costs. An adequate reactor design helps to avoid this critical drawback of ozonation, as will be discussed later.

Furthermore, if high levels of bromide ions are present in the effluent, ozonation can lead to the generation of bromate, which has been proved to act as a carcinogen [80].

The ozonation treatment can be improved by coupling with hydrogen peroxide and/or UV (O_3/UV, O_3/H_2O_2, or O_3/H_2O_2/UV processes). The next sections are devoted to these binary or ternary systems.

3.2. The O_3/UV System

Ozonation alone poses several advantages over conventional chemical oxidants such as chlorine or chlorine dioxide; however, it does not generate enough concentration of hydroxyl radicals to degrade organic compounds until total mineralization. This latter is due to the low value of the kinetic constant of the direct ozone-pollutant reaction. Hence, the concentration of hydroxyl radicals generated by decomposition of the ozone is insufficient, unless the pH of the medium rises, which would imply the use of external chemical agents. In short, higher concentrations of these radicals are required for the reduction of the pollutants.

The O_3/UV combination generates large concentrations of hydroxyl radicals in a fast manner so that this technique is adequate for this type of mineralization processes. Therefore, ozonation in the presence of UV irradiation has become one of the most used AOPs for the degradation of organic compounds in general as acids, alcohols, and organochlorines of low molecular weight (dihalomethanes, trihalomethanes, etc.). Another essential advantage of the combined use of UV and O_3 is the fact that the generation of bromate is inhibited.

The molar extinction coefficient of O_3 is 3300 $M^{-1}\cdot cm^{-1}$. Ozone strongly absorbs UV light of wavelength $\lambda = 254$ nm. To explain the generation of hydroxyl radicals from O_3, a two-stage process has been proposed. In the first stage, the photoinduced homolysis of the ozone molecule takes place, Reaction (20), and in the second stage, the production of hydroxyl radicals as a consequence of the reaction of atomic oxygen O (^{1}D) with water, Reaction (21), takes place [81]:

$$O_3 + h \rightarrow O_2 + O(^1D) \quad \text{(Reaction 20)}$$

$$O(^1D) + H_2O \rightarrow 2HO \quad \text{(Reaction 21)}$$

However, the hydroxyl radicals recombine generating hydrogen peroxide; therefore, the photolysis of ozone in solution can be represented by Reaction (22).

Hv

$$O_3 + H_2O \rightarrow [2 \cdot OH] + O_2 \rightarrow H_2O_2 + O_2 \quad \text{(Reaction 22)}$$

The H_2O_2 molecules generated in this latter reaction may undergo different chemical reactions in the presence of ozone as follows [82]:

$$H_2O_2 \rightarrow HO_2^- + H^+ \quad \text{(Reaction 23)}$$

$$HO_2^- + O_3 \rightarrow HO_2\cdot + O_3^- \quad \text{(Reaction 24)}$$

$$HO_2\cdot \rightarrow O_2^-\cdot + H^+ \quad \text{(Reaction 25)}$$

$$O_2^-\cdot + O_3 \rightarrow O_2 + O_3^- \quad \text{(Reaction 26)}$$

$$O_3^-\cdot + H^+ \rightarrow HO_3 \quad \text{(Reaction 27)}$$

$$HO_3\cdot \rightarrow HO\cdot + O_2 \quad \text{(Reaction 28)}$$

Next,

$$RH + OH \rightarrow H_2O + R \quad \text{(Reaction 1)}$$

$$R{\cdot} + O_2 \rightarrow ROO \quad \text{(Reaction 2)}$$

Peyton and Glaze [81] suggest that initiation may occur either by the reaction of ozone with the HO^- or HOO^- species or by the photolysis of hydrogen peroxide. The reaction with water of the radical anion ozonate, $O_3^{-}{\cdot}$is rapid, Reaction (29). The spontaneous decomposition of the ozonate has also been proposed [83], Reaction (30).

$$O^{-}{\cdot} + H_2O \rightarrow OH + HO^- \quad \text{(Reaction 29)}$$

$$O_3^{-}{\cdot} \rightarrow O_2 + O^- \quad \text{(Reaction 30)}$$

Hence, as described in Reaction (1), the hydroxyl radicals react with the organic substances, and the kinetics of the process is speeded up by the presence of UV radiation. Of course, as indicated in the previous section, UV light itself can degrade some compounds by direct photolysis. Moreover, UV radiation can excite the organic molecules of the pollutant, increasing their susceptibility towards an attack by the hydroxyl radicals.

Nevertheless, since both UV and ozone are quite expensive to generate and need the consumption of large amounts of electric energy, there are relatively few works in the literature devoted to the study of the removal of pollutants by O_3/UV processes in comparison with other UV- or ozone-based systems. Table 6 summarizes some research works dealing with the removal of several antibiotics by the O_3/UV process.

Table 6. Removal efficiency of antibiotics in waters by the O_3/UV process.

Antibiotic	Matrix	Operation Conditions	Maximum Removal Efficiency	Remarks	Reference
Amoxicillin (AMX)	Ultrapure water	O_3 flow:16 mg·h^{-1} T = 20 °C Low-pressure mercury vapor lamp (λ = 254 nm)	~100%	Synergistic effect of direct ozonation, direct photolysis, and hydroxyl radical oxidation. ·OH radicals generated in the photolysis of O_3	[84]
Azithromycin (AZT) Norfloxacin (NRF) Ofloxacin (OFX) Roxithromycin (RXT)	Ultrapure water WWTP effluent	0.08 mM O_3 T = 20 °C Low-pressure mercury vapor lamp (λ = 254 nm)	≥98% (water) >87% (effluent)	Synergetic effect between O_3 and UV irradiation	[85]
Azithromycin (AZT) Ciprofloxacin (CPR) Clarithromycin (CLR) Erythromycin (ERY) Levofloxacin (LVF) Lincomycin (LNC) Nalidixic acid (NLD) Roxithromycin (RXT) Sulfadimethoxine (SLM) Sulfamethoxazole (SLF) Trimethoprim (TRM)	WWTP effluent	Pilot-scale plant 0.02–0.12 mM O_3 Room temperature Low-pressure mercury vapor lamp (λ = 254 nm)	CPR, SLM, LNC, NLD, SLF, LVF, ERY, and TRM below detection limit (LOD). CLR, AZT, and RXT insensitive or very stable in the O_3/UV process	31 out of 38 PPCPs detected in the secondary effluent were degraded to or below their LOD	[86]
Ciprofloxacin (CPR) Trimethoprim (TRM),	Ultrapure water	2–20 mM O_3 Medium-pressure (MP) polychromatic UV lamp (λ=200–300 nm)	~100%	O_3-based processes more efficient than UV-based processes	[87]
Chloramphenicol (CHL)	groundwater surface water WWTP effluent	Low-pressure mercury vapor lamp (λ = 254 nm) pH = 8.0–8.2	>90%	Abatement efficiencies only moderately increased compared to conventional ozonation	[88]
Sulfamethoxazole (SLF)	Ultrapure water	0.03 mM O_3 High-pressure mercury lamp (λ = 313 nm) Room temperature	~100%	Complete removal achieved in 10 min	[89]

The literature review suggests that the use of UV light combined with O_3 increases the removal efficiency of refractory pollutants. This fact is attributable to the photolysis-mediated generation of larger amounts of ·OH radicals. However, the economic viability of this method is limited because of the remarkable energy requirement for the production of O_3 and UV light.

3.3. The O_3/H_2O_2 System

The direct action of ozone on certain typical water pollutants is advantageous, as discussed above, due to its ability to degrade high molecular weight electron-rich organic compounds. However, the decomposition of these large molecules usually gives rise to the generation of low molecular weight byproducts that may be refractory or recalcitrant towards both further ozone oxidation or degradation through ·OH radical-mediated pathways. Hence, it is possible that though the primary pollutants are entirely degraded by single ozonation, the degree of mineralization can be deficient [77]. This is, perhaps, the most critical disadvantage of the use of single ozonation as an AOP, since these low molecular weight byproducts may exhibit more acute toxicity than the primary pollutants. The concomitant use of hydrogen peroxide in ozonation may help to improve the process efficiency, since the reaction of O_3 with H_2O_2 results in the generation of ·OH radicals. The O_3/H_2O_2 combined oxidation system (commonly known as "peroxone") produces higher conversion yields than ozonation in those cases in which the direct ozone-pollutant reaction follows a slow kinetic regime due to gas–liquid matter transfer problems. Under these circumstances, an advanced oxidation process such as the one that can be achieved by adding a small amount of hydrogen peroxide to the aqueous solution through which ozone is being bubbled is convenient [90].

Hydrogen peroxide in aqueous solution is partially dissociated into its conjugate base, the hydroperoxide ion (HO_2^-), according to Reaction (23). Hydroperoxide ions react with ozone causing its decomposition, Reaction (24), and giving rise to a series of chain reactions in which the hydroxyl radical is involved [32]. Such reactions are much the same as indicated above as Reactions (25)–(28). Furthermore, ozone can react with hydroxyl radicals, Reaction (31), giving rise to more hydroperoxide ions that can further react with ozone, thus making the process continue. This way, the pollutant dissolved in water is susceptible to undergo oxidation through two simultaneous routes: The direct route (molecular reaction with ozone) or the indirect radical pathway (reaction with the hydroxyl radical).

$$H_2O_2 \rightarrow HO_2^- + H^+ \quad \text{(Reaction 23)}$$

$$HO_2^- + O_3 \rightarrow HO_2\cdot + O_3^- \quad \text{(Reaction 24)}$$

$$HO_2\cdot \rightarrow H^+ + O_2^- \quad \text{(Reaction 25)}$$

$$O_2^-\cdot + O_3 \rightarrow O_2 + O_3^- \quad \text{(Reaction 26)}$$

$$O_3^-\cdot + H^+ \rightarrow HO_3 \quad \text{(Reaction 27)}$$

$$HO_3\cdot \rightarrow HO\cdot + O_2 \quad \text{(Reaction 28)}$$

$$HO\cdot + O_3 \rightarrow HO_2\cdot + O_2 \quad \text{(Reaction 31)}$$

Additionally, ozone may also react directly with hydrogen peroxide and more hydroxyl and ozonate radicals are generated:

$$2\,H_2O_2 + O_3 \rightarrow 2\,HO\cdot + O_3^- \quad \text{(Reaction 32)}$$

Some selected results obtained in the O_3/H_2O_2 process are listed in Table 7.

Table 7. Removal efficiency of antibiotics in waters by the O_3/H_2O_2 process.

Antibiotic	Matrix	Operation Conditions	Maximum Removal Efficiency	Remarks	Reference
Amoxicillin (AMX)	Ultrapure water Reservoir water Groundwater, Two secondary WWTP effluents	O_3 flow: 16 mg·h^{-1} H_2O_2= 10 μM T = 20 °C	~100% in all cases	O_3/H_2O_2 process leads to the highest rate constants. Degradation rate higher in the UP. Dissolved organic matter results in slower degradation process	[84]
Ciprofloxacin (CPR)	Ultrapure water	O_3 = 0.1 mM H_2O_2 = 2–990 μM	95% degradation reached after 60–75 min	No effect of temperature (6.0–62.0 °C). Low [H_2O_2] (2–50 μM) increased CPR degradation Large concentrations (990 μM) decreased degradation rates at pH 7	[91]
Ciprofloxacin (CPR)	WWTP effluent	O_3 = 0.23 mM H_2O_2 = 20 mM 0.15 mL H_2O_2 (30% *w/v*) injected every 5 min	>99% after 5 min >99.5% after 10 min	High degree of mineralization (>90%)	[92]
Levofloxacin (LVF)	Ultrapure water	O_3 = 0.1 mM H_2O_2 = 2–100 μM pH = 3–10	95% (40 min) 99% (50 min)	Strong influence of pH on levofloxacin degradation rate and reaction pathways H_2O_2 addition had only a limited effect	[66]
Metronidazole (MTR)	WWTP effluent	O_3 = 0.23 mM H_2O_2 = 20 mM 0.15 mL H_2O_2 (30% *w/v*) injected every 5 min	>92% after 5 min	Low molecular weight carboxylates (mostly oxalates) as the final product	[92]
Sulfamethoxazole (SLF)	Ultrapure water Spiked WWTP effluent	O_3 = 0.42 mM H_2O_2 = 5 mM	~100%	Water matrix has no significant impact on SLF removal. Total degradation achieved in 45 min	[93]

Researchers agree that the O_3/H_2O_2 system is highly effective in achieving fast and complete mineralization of recalcitrant organic pollutants that can be found in wastewaters. The addition of H_2O_2 accelerates O_3 decomposition and the subsequent formation of ·OH radicals. This, in turn, makes the overall process faster since the reaction rate constant of the hydroxyl radical (i.e., 10^6–10^9 M^{-1} s^{-1}) is several orders of magnitude higher than that of O_3.

3.4. The $O_3/H_2O_2/UV$ System

In the $H_2O_2/O_3/UV$ system, there is a wide variety of individual processes that can give rise to the generation of ·OH radicals. Hence, this ternary system can be considered as the result of the integration of different unitary or binary systems, namely:

(a) Direct photolysis.
(b) Ozonation alone.
(c) UV photolysis of O_3.
(d) The combined effect of O_3 and H_2O_2.
(e) UV photolysis of H_2O_2.

The methods (a) to (d) have been considered in previous sections of this work, whereas the UV photolysis of hydrogen peroxide will be dealt with in depth in Section 4.1.

All these processes result in the formation of ·OH radicals through a sequence of reactions that directly or indirectly are related to the formation of (and/or the reaction with) H_2O_2 as suggested by reactions

$$O_3 + H_2O \xrightarrow{Hv} [2{\cdot}OH] + O_2 \rightarrow H_2O_2 + O_2 \quad \text{(Reaction 22)}$$

$$2\,H_2O_2 + O \rightarrow 2\,HO + O_3^-$$ (Reaction 32)

among others.

The main advantage of the ternary $O_3/H_2O_2/UV$ system lies in the fact that the decomposition of ozone is speeded up by the simultaneous presence of hydrogen peroxide and UV irradiation, thus yielding an increased rate of generation of ·OH radicals. Furthermore, it can also be applied under mild conditions (namely, atmospheric pressure and room temperature). However, the high costs of the three elements that constitute the system (i.e., ozone, hydrogen peroxide, and UV light) pose a remarkable disadvantage that limits a broader use of this process. Consequently, the use of this ternary system is usually restricted to the treatment of highly polluted effluents to achieve adequate degradation and mineralization of recalcitrant pollutants.

Some examples of the use of the $O_3/H_2O_2/UV$ system in the removal of antibiotics are summarized in Table 8.

Table 8. Removal efficiency of antibiotics in waters by the $O_3/H_2O_2/UV$ process.

Antibiotic	Matrix	Operation Conditions	Maximum Removal Efficiency	Remarks	Reference
Berberine (BRB)	Synthetic & real wastewater spiked with 1500 mg/L of BRB	H_2O_2 = 0.5–4.0 mM Low-pressure mercury vapor lamp (λ = 254 nm) pH = 5–11	94.1%	Performance of the process mainly relied on the H_2O_2 and O_3 dosages, water alkalinity, and contact time	[94]
Chlortetracycline (CHL)	Livestock wastewater	O_3 = 0.012 mM H_2O_2 = 0–5.9 mM Low-pressure mercury vapor lamp (λ = 254 nm) pH = 8.5	100% in less than 15 min	Complete mineralization not achieved	[95]
Ciprofloxacin (CPR) Trimethoprim (TRM)	2.5 mM phosphate buffer saline (PBS) at pH 7	O_3 = 0.1 mM H_2O_2 = 0.05–0.1 mM Medium-pressure (MP) polychromatic UV lamp (λ =200–300 nm)	>90%	Larger contribution of O_3-mediated degradation pathways (O_3 for TRM and O_3/H_2O_2 for CPR). UV contributes to a lesser extent	[87]
Penicillin G (PNG)	Ultrapure water	O_3 = 0.03 mM H_2O_2 = 3 mM Low-pressure mercury vapor lamp (λ = 254 nm)	~80% in 30 min	O_3 alone was very effective A complete degradation or mineralization was not achieved	[96]
Sulfamethoxazole (SLF)	Spiked ultrapure and tap water	O_3 = 0.04 mM H_2O_2 = 1 mM Low-pressure mercury vapor lamp (λ = 254 nm) pH = 3–10	100%	100% removal obtained in O_3/UV system	[97]

4. Hydrogen Peroxide-Based AOPs

Hydrogen peroxide is a safe, abundant, and easy to use chemical reagent, widely used for the prevention of contamination. However, H_2O_2 itself does not exhibit good oxidizing properties and must be combined with other substances or catalysts to become more effective. The combined use of H_2O_2 with ozone has been described under Section 3.2. Hydrogen peroxide can be used in combination with the UV radiation but has also been widely applied together with Fe^{2+} and/or Fe^{3+} ions giving rise to the well-known Fenton and Fenton-like processes. Finally, a ternary combination of UV irradiation, H_2O_2, and Fe^{2+}/Fe^{3+} ions, commonly known as the "Photo-Fenton process," has been broadly reported in the literature. In the next sections, these processes will be described in some detail.

4.1. The H_2O_2/UV System

Ultraviolet radiation has been widely used for the treatment of water and wastewater around the world and has more and more applications in this field. Numerous studies show that this treatment is useful for the elimination of pharmaceuticals found in different types of surface water [59,98,99]. However, this technology is only applicable to waters containing photosensitive compounds and with low levels of COD (for example, river and drinking water) [41]. On the other hand, effluents from sewage treatment plants may contain high concentrations of organic compounds that can inhibit the process [100]. In these cases, UV/H_2O_2 is a particularly attractive alternative for the removal of organic molecules that exhibit low reactivity towards ozone and hydroxyl radicals, but that are markedly photoactive.

The UV/H_2O_2 system takes advantage of the joint action of two chemical processes, namely

(a) The UV photolytic ability (regardless it is direct or indirect),
(b) The reaction of the dissolved pollutants with the ·OH radicals generated in the homolytic cleavage of the O-O bond in H_2O_2.

It can be stated that the photolysis of an organic compound in aqueous solution catalyzed by the presence of hydrogen peroxide is a very complex process. The success of the H_2O_2/UV system lies in the stoichiometric formation of hydroxyl radicals (·OH) from the photocatalytic decomposition of H_2O_2 in the first stage of the photolytic degradation [101]:

Hv

$$H_2O_2 \rightarrow 2\ OH \quad \text{(Reaction 33)}$$

The quantum yield of this process is very high, forming a maximum of two hydroxyl radicals as absorbed, and invariable with the applied wavelength [102].

Next, a series of radical reactions takes place:

$$H_2O_2 + OH \rightarrow HO_2\cdot + H_2O \quad \text{(Reaction 34)}$$

$$H_2O_2 + HO_2\cdot \rightarrow OH + O_2 + H_2O \quad \text{(Reaction 35)}$$

$$2\ HO_2\cdot \rightarrow H_2O_2 + O_2 \quad \text{(Reaction 36)}$$

$$OH + HO_2\cdot \rightarrow H_2O_2 + O_2 \quad \text{(Reaction 37)}$$

From Reactions (34)–(37) it becomes evident that although the photolytic cleavage of hydrogen peroxide gives rise to the formation of ·OH radicals (Reaction (33)), the occurrence of high concentrations of H_2O_2 may have a scavenging effect on the hydroxyl radicals and, hence, may hinder the effectiveness of the oxidation process. Consequently, the H_2O_2 initial concentration must be carefully adjusted to maximize the efficiency of the removal process. Moreover, H_2O_2 is an expensive reagent that increases the total operating costs of the process [103].

Once the highly reactive ·OH radicals are formed, they react with the organic compound by different mechanisms: Abstraction of a hydrogen atom, addition to C=C double bonds, or electron transfer, depending on the nature and functional groups of the molecule. The most general reaction route is the abstraction of a hydrogen atom and the generation of the resulting organic radical R·, which in turn reacts rapidly with dissolved O_2 to form the peroxide organic radical $RO_2\cdot$ [21]. These organic radicals decompose by bimolecular reactions giving rise to the different degradation products of the starting compound together with other byproducts such as hydrogen peroxide, hydroperoxide radicals, formaldehyde, etc.

Finally, the dimerization reactions of the hydroxyl radicals themselves, i.e., the reverse process of Reaction (33) [104] and the hydroperoxide radicals, Reaction (36) [105], lead to the regeneration of hydrogen peroxide, which in turn can sequester hydroxyl radicals and re-form hydroperoxide radicals, Reaction (34) [106].

At the same time, one must consider the dissociation equilibria of the organic compound itself and of the different intermediates formed, such as hydrogen peroxide, hydroperoxide radicals, etc., which are shown below:

$$RH \leftrightarrow R^- + H^+ \quad \text{(Reaction 38)}$$

$$H_2O_2 \rightarrow HO_2^- + H^+ \quad \text{(Reaction 23)}$$

$$HO_2\cdot \rightarrow H^+ + O_2^- \quad \text{(Reaction 25)}$$

In summary, a cycle of decomposition and simultaneous formation of hydrogen peroxide can be established. The overall result of such a cycle will depend on several variables as, for instance, the intensity of the ultraviolet radiation, temperature, pH, and the nature of the organic compounds.

It must be noted, however, that the H_2O_2/UV system is considered one of the most viable advanced oxidation processes. For instance, on many occasions, it is preferable to ozonation because it is less sensitive to the nature and concentration of the polluting species.

Table 9 summarizes some examples of removal processes of antibiotics by the H_2O_2/UV process.

From the literature review, it may be concluded that the UV/H_2O_2 is a fast and efficient technology for the removal of antibiotics from aqueous matrices, due to a fast generation of OH radicals in solution. However, the overall performance of the process is remarkably dependent on different operational parameters such as the UV wavelength and intensity (i.e., the UV light source) and the inherent properties of the wastewater (i.e., pH, initial concentration of pollutants, etc.).

Table 9. Removal efficiency of antibiotics in waters by the H_2O_2/UV process.

Antibiotic	Matrix	Operation Conditions	Maximum Removal Efficiency	Remarks	Reference
Amoxicillin (AMX)	Ultrapure water	H_2O_2 = 0.4–10 mM Low-pressure mercury lamp (λ = 254 nm) pH = 2–10	99% in 20 min	Low mineralization Antibacterial activity effectively eliminated	[107].
Cefalexin (CFL) Norfloxacin (NRF) Ofloxacin (OFX)	Ultrapure water and tap water	H_2O_2 = 0.25–5.0 mM Low-pressure mercury vapor lamp (λ = 254 nm)	~100% within 3–5 min	Scavenging effect if large concentrations of H_2O_2 are used Presence of halides in tap water accelerates the degradation rate	[108]
Ceftriaxone (CFN)	Ultrapure water and tap water	H_2O_2 = 0.15–2.9 mM Low-pressure mercury vapor lamp (λ = 254 nm) pH = 5–9	~100%	Optimum removal efficiency reached for [H_2O_2] = 0.3 mM Up to 35% synergistic effect achieved with respect to the photolysis process 58.1% mineralization reached	[109]
Ciprofloxacin (CPR) Doxycycline (DXY) Oxytetracycline (OXY)	Ultrapure water Surface water Wastewater	H_2O_2 = 0.7–4.2 mM Low-pressure mercury lamp (λ = 254 nm) pH = 2–10	100%	Toxicity firstly increases, then decreases 10% mineralization, total mineralization needed much more energy	[99]
Norfloxacin (NRF)		H_2O_2 = 0.7–4.2 mM Low-pressure mercury lamp (λ = 254 nm) pH = 2–10	100% in 100 min	Direct photolysis infeasible (high reaction time and low mineralization)	[110]
Ofloxacin (OFX) Sulfaquinoxaline(SLQ)	Ultrapure water	H_2O_2 = 0.8–9.0 mM Low-pressure mercury lamp (λ = 254 nm)	>99% in 11 min	OFX is degraded faster than SLQ Degradation products of OFX and SLQ are harmful to microorganisms	[111]
Roxithromycin (RXT)	Ultrapure water Secondary wastewater effluent	H_2O_2 = 2–20 mM High-pressure mercury lamp (λ = 365 nm) pH = 4–9	~100% in appr 45 min	Slightly alkaline favorable for the RXT degradation Degradation products more toxic than the parent compound	[112]
Sulfadiazine (SDZ) Sulfathiazole (STZ) Sulfamerazine (SMR) Sulfisoxazole (SSX) Sulfamethazine (SMZ) Sulfamethoxypyridazine (SMP) Sulfamonomethoxine (SMM) Sulfadimethoxypyrimidine (SDM)	Synthetic wastewater Hydrolyzed urine 5 mM phosphate buffer (pH = 7)	H_2O_2 = 0.9 mM Low-pressure mercury lamp (λ = 254 nm)	>99%	Sulfonamides with five-membered heterocyclic group undergo rapid direct photolysis.	[113]
Sulfamethazine (SMZ)	Ultrapure water	H_2O_2 = 1–10 mM Low-pressure mercury lamp (λ = 254 nm)	100% in 10 min	57% mineralization in 120 min	[114]
Sulfamethoxazole (SLF)	Ultrapure water	H_2O_2 = 0, 0.15, and 0.30 mM Low-pressure and medium-pressure mercury lamps (λ = 254 and 365 nm)	~100%	Removal largely attributed to direct photodegradation Lower UV or H_2O_2 doses yielded different relative abundances of certain transformation products as compared to higher doses	[115]

4.2. The Fe^{2+}/H_2O_2 System. Fenton Reagent

The Fenton process is a viable alternative for the removal of organic pollutants from wastewater and has been applied in many industrial sectors. However, it has some disadvantages derived from the use of iron salts as a catalyst for the decomposition of hydrogen peroxide to hydroxyl radicals.

On the one hand, large amounts of these dissolved iron salts are necessary, which makes the process more expensive. On the other hand, the directives of the European Union on water quality allow a very low concentration of iron dissolved in the effluents, which forces to introduce some treatment aimed at eliminating iron salts from the effluents of the Fenton process. These complementary processes, typically physical–chemical coagulation–flocculation processes, produce large quantities of metal sludge as a waste, which must be managed appropriately too. These drawbacks of the conventional Fenton process have promoted the development of new systems, which allow minimizing the presence of iron species dissolved in the environment, without critically affecting the efficiency of the process.

Different alternatives can be distinguished to achieve this objective [116]. Firstly, technological options have been proposed to accelerate the regeneration of Fe(II) species, which is mainly responsible for the decomposition of hydrogen peroxide and the generation of hydroxyl radicals. This would reduce the amount of iron (II) needed in the reaction medium. One of these alternatives is the combined use of the Fenton process together with near-visible ultraviolet radiation, which constitutes the so-called "photo-Fenton process." This process will be dealt with in detail in Section 4.4. Secondly, the development of solid catalysts for Fenton processes has attracted the attention of the scientific community in recent years. The use of highly active and stable solid catalysts would help to minimize the presence of iron (II) salts dissolved in the medium, besides facilitating the recovery and reuse of the catalyst. This latter alternative gives rise to a broad group of process commonly known as "heterogeneous Fenton processes." Finally, the combined use of Fenton's reagent and electric current receives the generic denomination of "electro-Fenton techniques" [117].

The addition of iron salts as a catalyst in the presence of hydrogen peroxide is one of the classical methods of producing hydroxyl radicals, being one of the most powerful oxidizing agents at acidic pH (namely, pH = 3–5). The Fenton reagent has a great oxidizing capacity towards a wide range of organic substances, both aromatic (phenols, polyphenols, etc.) and aliphatic compounds (alcohols, aldehydes, etc.). The main oxidizing species is again the hydroxyl radical, which is generated in the initial reaction between hydrogen peroxide and Fe^{2+} salts [118]. While the exact mechanism of the oxidation of an organic compound by the reagent mentioned above is complex and not completely known, several authors agree on its main stages. Thus, it can be assumed that the overall process takes place through the following individual stages:

$$Fe^{2+} + H_2O_2 \rightarrow Fe^{3+} + OH + OH^- \qquad \text{(Reaction 39)}$$

$$R + H_2O_2 \rightarrow P1 \qquad \text{(Reaction 40)}$$

$$R + \cdot OH \rightarrow P2 \qquad \text{(Reaction 41)}$$

where R represents the organic compound, and P1 and P2 are the formed intermediates and final products of the oxidation. The first reaction of the mechanism is responsible for the formation of hydroxyl radicals [118] that will later attack organic compound R in Reaction (41), the main degradation pathway in the Fenton reaction.

On the other hand, Reactions (34) and (42) represent the sequestering effect of such radicals exerted by Fe^{2+} itself or H_2O_2 [119]:

$$H_2O_2 + OH \rightarrow HO_2 + H_2O \qquad \text{(Reaction 34)}$$

$$Fe^{2+} + OH \rightarrow Fe^{3+} + OH^- \qquad \text{(Reaction 42)}$$

Finally, reactions (43), (44), and (45) indicate the possible reaction pathways of the Fe(III) generated in reaction (42) with H_2O_2 and with the hydroperoxide ($HO_2\cdot$) or superoxide ($O_2^{-}\cdot$) radicals.

$$Fe^{3+} + H_2O_2 \rightarrow Fe^{2+} + H^+ + HO_2 \quad \text{(Reaction 43)}$$

$$Fe^{3+} + HO_2\cdot \rightarrow Fe^{2+} + H^+ + O_2 \quad \text{(Reaction 44)}$$

$$Fe^{3+} + O_2^{-}\cdot \rightarrow Fe^{2+} + O_2 \quad \text{(Reaction 45)}$$

It is worth noting that in Reactions (43)–(45), Fe^{2+} is regenerated, so that the Fenton process can be regarded as catalytic with respect to iron. Therefore, the reaction of formation of OH radicals can continue to take place as long as there is hydrogen peroxide in the medium.

For the procedure to be effective, the following requirements are necessary:

(a) The pH of the water to be treated must be in the range 3–5 since at higher pH values, iron precipitates as $Fe(OH)_3$, thus inactivating the system. Furthermore, if the pH is high, the decomposition of hydrogen peroxide into oxygen and water is favored.
(b) The addition of the Fe^{2+} salt is necessary, generally as $FeSO_4$, even though other sources of Fe^{2+} or Fe^{3+} can be added. In the case of Fe^{3+}, which is also useful, a small initial delay of the reaction is observed.
(c) The addition of H_2O_2 must be very slow to avoid decomposition phenomena.

The rate of ·OH radical generation, which in turn depends on the concentration of ferrous catalyst, generally limits the reaction rate of this system. The typical Fe^{2+}: H_2O_2 molar ratio is 1: 5–10, although Fe^{2+} levels below 25–50 $mg{\cdot}L^{-1}$ may require a considerable reaction time (10–48 h).

The main advantages of this oxidation process are:

(a) Fe^{2+} is abundant and non-toxic.
(b) Hydrogen peroxide is easy to handle and environmentally benign.
(c) No chlorinated compounds are formed as in other techniques.
(d) There are no limitations of matter transfer since the system is homogeneous. Hence, the design of reactors for this technology is quite simple.
(e) An additional advantage of the Fenton process is the formation of complexes that promote the coagulation of suspended solids after oxidation reactions [120].

However, the Fenton process has some shortcomings, including:

(a) The regeneration rate of Fe^{2+} from Fe^{3+} according to Reactions (43)–(45) is very low if compared with the depletion rate of Fe^{2+} (Reaction (42))
(b) If pH increases above 3 or 3.5, large amounts of sludge are produced due to iron hydroxide precipitation, and additional treatment is necessary.
(c) Fe^{2+} or Fe^{3+} ions may undergo complexation reactions with organic or inorganic ligands that may be present in solution.
(d) Scavenging reactions may take place, for instance, Reaction (42).

In the absence of ferrous or ferric salt, there is no evidence of hydroxyl radical formation. As the iron concentration increases, the oxidation rate of organic compounds increases to a point at which an additional increase in iron concentration is ineffective. For most applications, it does not matter if Fe^{2+} or Fe^{3+} is used; the catalyst cycle starts quickly if hydrogen peroxide and organic material are in sufficient concentration.

When the H_2O_2 dose is increased, a noticeable reduction in organic matter is obtained, whereas a small or negligible change in toxicity may occur. Once a minimum threshold has been reached, small increases in the H_2O_2 dose result in evident decreases in the toxicity of the effluent. However, it must be kept in one's mind that high concentrations of hydrogen peroxide lead to the scavenging of ·OH radicals (see Reactions (34)–(37)).

The reaction rate in the Fenton process increases with temperature, the effect being more pronounced at temperatures below 20 °C. However, when the temperature rises to 40–50 °C, the effectiveness of the reagent decreases. This is due to the accelerated decomposition of H_2O_2 into oxygen and water. From a practical standpoint, most of the commercial applications of this reagent occur at a temperature between 20–40 °C.

As indicated above, the optimum operational pH is between 3 and 3.5. The inefficiency of a basic pH is attributed to the transformation of the hydrated iron species to colloidal ferric species. In this last form, iron catalytically decomposes hydrogen peroxide into oxygen and water, without forming hydroxyl radicals.

This process can be applied to wastewater, sludge, or contaminated soils producing the oxidation of organic pollutants, reduction of toxicity, reduction of COD, reduction of BOD_5, and elimination of odor and color. Some examples of the use of the Fenton process are summarized in Table 10.

Table 10. Removal efficiency of antibiotics in waters by the Fenton process.

Antibiotic	Matrix	Operation Conditions	Maximum Removal Efficiency	Remarks	Reference
Amoxicillin (AMX)	Spiked wastewater	H_2O_2 = 0.3–15 mM Fe^{2+} = 0–0.9 mM	100% in 2.5 min	Box-Behnken-statistical design 37% mineralization in 15 min	[121]
Amoxicillin (AMX)	Ultrapure water	H_2O_2 = 0.1–0.125 mM Fe^{2+} = 0.004–0.006 mM pH = 3.5 T = 20–30 °C	100% in 30 min	Central composite factorial design Only T and [Fe^{2+}] affect statistically the removal efficiency	[122]
Amoxicillin (AMX)	Synthetic wastewater	H_2O_2 = 5–50 mM Fe^{2+} = 0.5–5 mM pH = 2–7	83%	Optimum Fe^{2+}/H_2O_2 molar ratio = 1/15 66% mineralization	[123]
Ampicillin (AMP)	Ultrapure water	H_2O_2 = 0.373 mM Fe^{2+} = 0.087 mM pH = 3.5	100% in 10 min	Central composite factorial experimental design Degradation products without antibacterial activity	[124]
Chlortetracycline (CHL)	Wastewater	H_2O_2 = 0.3 mM Fe^{2+} = 0.003–0.3 mM pH = 7	76%	Complete mineralization not achieved	[125]
Ciprofloxacin (CPR)	Ultrapure water	H_2O_2 = 20–84 mM Fe^{2+} = 5–21 mM pH = 5	74% in 25 min	Optimal conditions: H_2O_2 = 74.5 mM Fe^{2+} = 17.46 mM pH = 4.6 Hydroxylation of both piperazine and quinolone rings; oxidation and cleavage of the piperazine ring, and **defluorination** (OH/F substitution) are the main degradation mechanisms	[126]
Ciprofloxacin (CPR)	Ultrapure water	H_2O_2 = 26–51 mM Fe^{2+} = 5–10 mM pH = 3	76% in 45 min	Optimal conditions: H_2O_2 = 26 mM Fe^{2+} = 5 mM pH = 3 Complete mineralization could not be achieved	[127]
Ciprofloxacin (CPR)	Spiked wastewater	H_2O_2 = 14.2 mM Fe^{2+} = 0.284–2.84 mM pH = 3	70% in 15 min	55% mineralization achieved Considerable reduction in toxicity of the treated wastewater	[128]
Doxycycline (DXY)	Ultrapure water	H_2O_2 = 2.9–26.5 mM Fe^{2+} = 0.09–2.1 mM T = 0–40 °C pH = 5	100% in 10 min	Optimal conditions: H_2O_2 = 18 mM Fe^{2+} = 0.44 mM T = 35 °C	[129]
Flumequine (FLM)	Ultrapure water	H_2O_2 = 0.5–10 mM Fe^{2+} = 0.25–1 mM pH = 2.8	40% in 15 min	Low mineralization degree Deactivation of antimicrobial activity	[130]
Levofloxacine (LVF)	Ultrapure water	H_2O_2 = 0.375–1.5 mM Fe^{2+} = 0.0375–0.15 mM pH = 3	100%	Total removal achieved within 5–90 min according to experimental conditions **Defluorination, piperazinyl substituent transformation, and quinolone moiety modifications are the main degradation pathways**	[131]
Sulfamethoxazole (SLF)	Ultrapure water Synthetic wastewater	H_2O_2 = 0.5–4 mM Fe^{2+} = 0.025–0.2 mM pH = 3	100% in 10 min (ultrapure water) 53% in 30 min (synthetic wastewater)	Wastewater components had a negative effect on sulfamethoxazole degradation Degradation pathways: (a) Oxidation of $-NH_2$ in benzene ring by ·OH radicals followed by hydroxylation (b) -SH-Ph bond cleavage	[132]
Trimethoprim (TRM)	Ultrapure water Synthetic wastewater	H_2O_2 = 0.5–4 mM Fe^{2+} = 0.025–0.2 mM pH = 3	100% in ultrapure water 36% in synthetic wastewater	**Hydroxylation is the first degradation step, followed by the** cleavage of the C-C bond between the pyrimidine and the benzene rings	[133]

4.3. The Fe^{3+}/H_2O_2 System. Fenton-Like Reagent

The main drawback of using the Fenton system described above is the cost of the reagents, namely H_2O_2 and Fe^{2+}. For this reason, several methods have been developed to substitute Fe^{2+} with Fe^{3+} salts, whose price is lower than that of the Fe^{2+} salts.

Originally, the "Fenton-like" term was used in reference to a similar process to that described in the previous section, with the only difference that the reagent used is a mixture of Fe^{3+} and H_2O_2 where the hydrogen peroxide decomposes into hydroxyl radicals, and the Fe(III) is reduced to Fe(II) following the reaction:

$$Fe^{3+} + H_2O_2 \rightarrow Fe^{2+} + OH^- + OH \qquad \text{(Reaction 46)}$$

Several studies have shown that the decomposition rate of H_2O_2 and the oxidation rate of organic solutes are markedly slower using Fe^{3+}/H_2O_2 than Fe^{2+}/H_2O_2, with an optimal being achieved at pH = 3 [134,135].

Additionally, the Fenton-like process produces peroxyl radicals ($HO_2\cdot$):

$$Fe^{3+} + H_2O_2 \rightarrow H^+ + FeOOH^{2+} \qquad \text{(Reaction 47)}$$

$$FeOOH^{2+} \rightarrow HO_2\cdot + Fe^{2+} \qquad \text{(Reaction 48)}$$

Despite the fact that the homogeneous Fenton or Fenton-like processes have been largely used due to their effectiveness in terms of pollutant removal as well as to their ease of operation, both of them exhibit important disadvantages such as excessive sludge production and limited range of operational pH (usually below 3). Heterogeneous Fenton oxidation was developed to overcome these problems. In heterogeneous Fenton oxidation, a reaction takes place between hydrogen peroxide and Fe(III) in different forms, e.g., Fe_2O_3 or α-FeOOH, among others. If solid catalysts are used, in addition to the chemical reactions described above, physical adsorption occurs at the surface of the solid catalyst, which reduces sludge generation.

For all the exposure, the Fenton-like system is becoming progressively less used in recent years, and the number of manuscripts published is decreasing. On the contrary, heterogeneous Fenton-like processes, as well as those using different precursors to generate Fe(II) or Fe(III) ions in solution, are gaining importance. Hence, in this section, results corresponding not only to the Fenton-like process in its traditional sense, but also to some new alternatives will be presented.

Fe(II) used in the traditional Fenton process can be efficiently substituted by nanoscale zero-valent iron (nZVI) that is commonly synthesized by the reaction of Fe(II) with sodium borohydride:

$$Fe(H_2O)_6{}^{2+} + 2\,BH_4{}^- \rightarrow Fe^0\downarrow + 2\,B(OH)_3 + 7\,H_2\uparrow \qquad \text{(Reaction 49)}$$

Once synthesized and/or isolated, Fe^0 reacts with hydrogen peroxide or dissolved oxygen in the acidic medium required for the Fenton process (pH ~ 2.5–3.5) yielding Fe^{2+} as follows:

$$Fe^0 + H_2O_2 + 2\,H^+ \rightarrow Fe^{2+} + 2H_2O \qquad \text{(Reaction 50)}$$

$$Fe^0 + O_2 + 2\,H^+ \rightarrow Fe^{2+} + H_2O_2 \qquad \text{(Reaction 51)}$$

Fe(II) undergoes the series of reactions described in the previous section. It is worth noting that Reaction (50) involves the degradation of hydrogen peroxide, whereas in Reaction (51), H_2O_2 is generated. Hence, the global reaction is:

$$2\,Fe^0 + O_2 + 4\,H^+ \rightarrow 2\,Fe^{2+} + 2\,H_2O \qquad \text{(Reaction 52)}$$

However, the heterogeneous process appears to be less effective than a homogeneous Fenton process due to mass-transfer limitation. To solve this problem, different metal oxides, MO_x (e.g., ceria [136], Fe_3O_4 [137], Mn_3O_4 [138], WMoO [139], $FeCuO_2$, $NiCuO_2$ [140], etc.) have been recently tested as catalysts or as a support of Fe^0 in a series of alternative Fenton-like processes. The use of these oxides involves either a faster kinetic removal or the broadening of the operational pH range, that may reach up to 10 in some of these processes, which is not suitable for conventional Fenton or Fenton-like processes due to Fe(II) and/or Fe(III) precipitation in the form of hydroxides or oxyhydroxides.

The improvement in the performance of the MO_x/Fe^0-catalyzed systems is attributable to different factors, mainly:

(a) MO_x usually exhibit larger specific surface areas than iron-based materials, thus favoring the adsorption of pollutants in the active sites of the solid's surface.
(b) MO_x may act as catalysts for the homolytic break of H_2O_2 into two ·OH radicals.
(c) MO_x used to have a relatively large number of oxygen vacancies that are suitable for pollutants to react rapidly with the reactive oxygen species (ROS) generated on the material's surface.

Biochar has also been used as catalyst support in Fenton-like processes [141]. It is well-known that the surface of biochars possesses a vast number and variety of redox-active sites (e.g., quinone, hydroquinone, conjugated π-electron systems, aromatic rings, etc.). These active sites are suitable to act as electron donors or acceptors in many redox processes such as Fenton-like, among others. Further details can be found in the excellent review recently published by Wang et al. (see reference [141] and citations therein).

Finally, the outstanding properties of graphene and graphene oxide have also been applied to the heterocatalytic Fenton-like process [142].

Table 11 summarizes some recent contributions regarding the removal of antibiotics through heterogeneous Fenton-like process

Table 11. Removal efficiency of antibiotics in waters through heterogeneous Fenton-like process.

Antibiotic	Matrix	Operation Conditions	Maximum Removal Efficiency	Remarks	Reference
Amoxicilin (AMX)	Ultrapure water	H_2O_2 = 3.3–12.2 mM nZVI = 0.5–2.0 g/L pH = 2.0–5.0 T = 15–45 °C	~90% in 20 min	(a)Optimal conditions: H_2O_2 = 6.6 mM, nZVI = 0.5 g/L pH = 3.0 T = 30 °C Adsorption of AMX onto nZVI or its (hydr)oxide surface plays an important role	[143]
Ciprofloxacin (CPR)	Ultrapure water	H_2O_2 = 100 mM nZVI = 0.056–0.28 g/L pH = 7 Room temperature	100% in 30 min	(a)Reaction at the piperazinyl ring and defluorination followed by hydroxyl substitution appear to be the main degradation pathways	[144]
Ciprofloxacin (CPR)	Ultrapure water	H_2O_2 = 10 mM Catalysts (0.5 g/L): $MnCuO_2$ $FeCuO_2$ $CoCuO_2$ $NiCuO_2$ pH = 6	~90%	(b)CPR degradation mainly occurs in solution. Scarce contribution of adsorption Degradation of CPR should be due to the cleavage of piperazine ring, followed by loss of formaldehyde, replacement of F with OH and/or loss of ethylamine	[140]
Ciprofloxacin (CPR)	Ultrapure water	H_2O_2 = 10–100 mM Sludge Biochar Catalyst (SBC) = 0.2 g/L pH = 2–12	90% in 4 h	(c)Fe^{2+} and Fe^{3+} were released in the SBC/H_2O_2 system Piperazine ring cleavage, pyridine cleavage, hydroxylation, F/OH substitution, and defluorination were the dominant degradation pathways	[145]
Metronidazol (MTR)	Ultrapure water	nZVI = 0.03–0.13 g L^{-1} pH = 3.03–9.04	96.4% in 5 min	(a)H_2O_2 generated according to Reaction (51)	[146]
Metronidazole (MTR)	Ultrapure water	Absence of Fe and H_2O_2 Addition of 2 mM H_2O_2 in one experiment	92%	(d)Three-dimensional macroporous graphene-wrapped zero-valent copper nanoparticles (3D-GN@Cu^0) used as the catalyst H_2O_2 generated in situ by reduction of O_2 on the surface of 3D-GN@Cu^0 Addition of 2 mM H_2O_2 had little effect on the degradation of MTR	[142]

Table 11. *Cont.*

Antibiotic	Matrix	Operation Conditions	Maximum Removal Efficiency	Remarks	Reference
Tetracycline (TTR)	Ultrapure water	H_2O_2 = 100 mM $Fe^0@CeO_2$ catalyst = 0.1 g/L pH = 5.8 T = 26 °C	94%	[(b)] A combined adsorption/reduction mechanism enhances removal efficiency	[136]
Tetracycline (TTR)	Ultrapure water	H_2O_2 = 3–20 mM Fe(III) concentration not specified $WMoO_x$ catalyst = 0.2–1.6 g/L pH = 3–8 T = 25 °C	86%	[(b)] The system avoids solution chroma and sludge formation caused by the dissolved ferric species	[139]
Tetracycline (TTR)	Ultrapure water Spiked wastewater	H_2O_2 = 5 mM pH = 7.4 T = 25 °C Fe substituted by a mixture of three biochars from corn stalks, bamboo, and pig manure	100%	[(c)] Pig manure showed the best performance in TTR removal	[147]
Tetracycline (TTR)	Ultrapure water	H_2O_2 = 1.1–3.3 mM α-FeOOH/RGO hydrogels used as catalysts	100% in 120 min	[(d)] α-FeOOH/RGO hydrogels could generate reactive oxygen species (ROS) without the addition of H_2O_2 TTR acts as an electron donor. e^- are transferred through π–π interactions (TTR -graphene) and π–Fe interactions (graphene- FeOOH)	[148]

KEY: Catalysts used: [(a)] nZVI; [(b)] MO_x; [(c)] Biochars; [(d)] Graphene/graphene oxide.

4.4. *The $Fe^{2+}/H_2O_2/UV$ System (Photo- Fenton).*

It is commonly accepted that UV radiation accelerates Fenton reactions, thus favoring the degree of degradation of organic pollutants, including aromatic and aliphatic compounds, and presenting greater effectiveness at acidic pH. The photo-Fenton system, therefore, includes ultraviolet radiation, hydrogen peroxide, and iron salts. This system has been considered one of the most promising ways of purifying highly contaminated wastewater [149,150].

The main advantages of the photo-Fenton process over the Fenton or Fenton-like reagents discussed in the previous section are the following:

(a) The photolysis of hydrogen peroxide, produced according to Reaction (33), provides a supplementary source of ·OH radicals [101].

$$H_2O_2 \xrightarrow{Hv} 2\,OH \qquad \text{(Reaction 33)}$$

(b) The reduction of Fe^{3+} to Fe^{2+} by ultraviolet radiation, shown in Reaction (53), also contributes to the generation of hydroxyl radicals. Furthermore, this reaction facilitates the formation of Fe(II), which reacts rapidly with hydrogen peroxide to yield more ·OH radicals by the conventional Fenton reaction (39). Hence, it can be stated that UV radiation accelerates the Fe(III)–Fe(II) cycle, thus facilitating the production of hydroxyl radicals in both reactions (53) and (39)

$$Fe(OH)^{2+} + h\nu \rightarrow Fe^{2+} + OH \qquad \text{(Reaction 53)}$$

$$Fe^{2+} + H_2O_2 \rightarrow Fe^{3+} + \cdot OH + OH^- \qquad \text{(Reaction 39)}$$

In addition to these important advantages, Bossman et al. [151] proposed that Fe(III) in the presence of ultraviolet radiation is promoted into an excited state of Fe(III) that reacts faster with hydrogen peroxide to form Fe(II) and OH radicals, Reaction (46), or even with organic compounds.

$$Fe^{3+} + H_2O_2 \rightarrow 2\,Fe^{2+} + OH^- + OH \qquad \text{(Reaction 46)}$$

All these advantages make the photo-Fenton system a promising procedure in the treatment of water purification, since hydroxyl radicals are generated and secondary chlorinated oxidation products are not produced, as in the case of oxidation by chlorine or chlorine dioxide.

Table 12 summarizes some recent papers reporting the use of the photo-Fenton process for the removal of pharmaceuticals from water.

Table 12. Removal efficiency of antibiotics in waters by photo-Fenton processes.

Antibiotic	Matrix	Operation Conditions	Maximum Removal Efficiency	Remarks	Reference
Amoxicillin (AMX)	Spiked synthetic wastewater Spiked real wastewater	H_2O_2 = 0.08 mM Fe^{3+} = 0.05 mM Natural solar radiation (pilot-plant scale CPC photoreactor) pH = 7–8	90% in 9 min	No mineralization of the drug. Hydroxylation of the aromatic ring, opening of the β-lactam ring, and subsequent formation of amoxilloic and amoxicilloic acids are the main transformation pathways.	[152]
Ampicillin (AMP)	Spiked WWTP effluent	Solar photo-Fenton H_2O_2 = 0.74–2.94 mM Fe^{2+} = 0.09 mM pH = 3	100% in 20 min	Optimal conditions: H_2O_2 = 2.2 mM Fe^{2+} = 0.09 mM Beyond the optimal H_2O_2 concentration, scavenging effects might occur	[153]
Chloramphenicol (CHL)	Spiked WWTP effluent	H_2O_2 = 0.044–0.088 mM Fe^{2+} = 0.016–0.064 mM Black light lamps (λ = 350–400 nm) or solar irradiation pH = 5.8–7.7	79% in 20 min	Optimal conditions: H_2O_2 = 0.088 mM Fe^{2+} = 0.048 mM pH = 5.8 Acidification and neutralization before the discharge are avoided	[154]
Ciprofloxacin (CPR)	Ultrapure water	H_2O_2 = 5–25 mM High-pressure mercury lamp (λ = 362 nm) T = 25 °C Fe^{2+} = 0.25–2 mM pH = 2–9	93% in 45 min	Optimal conditions: H_2O_2 = 10 mM Fe^{2+} = 1.25 mM pH = 3.5 70% mineralization reached Piperazine ring degradation is the main degradation pathway	[155]
Oxacillin (OXC)	Ultrapure water	H_2O_2 = 0.09–10 mM Fe^{2+} = 0.0036–0.09 mM High-pressure mercury lamp (λ = 365 nm) pH = 6 T = 25 °C	100% in 20 min	Optimal conditions H_2O_2 = 10 mM Fe^{2+} = 0.09 mM Light intensity = 30 W Effluent has no antimicrobial activity Near-neutral conditions are used	[156]
Trimethoprim (TRM)	Ultrapure water	H_2O_2 = 0.03–5 mM High-pressure mercury lamp (λ = 360 nm) T = 25 °C Fe^{2+} = 0.03–2 mM pH = 2.5–4.5	99.5% in 6 min	Optimal conditions: H_2O_2 = 0.09 mM Fe^{2+} = 0.09 mM pH = 4.56	[157]

According to the literature, the photo-Fenton process is economical, technically simple, and highly efficient for the removal of pollutants in general and antibiotics in particular from wastewaters. Fe(II) salts, the H_2SO_4 required for acidic pH, and hydrogen peroxide are readily available chemicals. Furthermore, the use of UV radiation speeds up the generation of ·OH radicals, thus reducing the H_2O_2 consumption in comparison with the traditional Fenton process.

5. Heterogeneous Photocatalysis with TiO_2

Photocatalysis is defined as the acceleration of a photochemical reaction by the presence of a semiconductor that is activated by the absorption of radiation with energy above its bandgap. The term *heterogeneous* refers to the fact that the contaminants are present in a fluid phase while the catalyst is in the solid phase. The most commonly used catalyst is titanium dioxide (TiO_2) due to its high chemical stability, low cost, and excellent results it has proven to provide [158,159].

The first reaction of the photocatalysis process is the absorption of UV radiation by the catalyst, with the formation of hollow-electron (h^+/e^-) pairs according to Reaction (54). In environmental applications, the photocatalytic processes are carried out under aerobic conditions, and oxygen can be adsorbed onto the catalyst's surface. Hence, the aforementioned electrons, due to their high reducing power, reduce the oxygen adsorbed on the TiO_2 surface, thus giving rise to the generation of superoxide radical ion $(O_2^-)\cdot$as indicated in Reaction (55); conversely, the holes are capable of causing the oxidation of water and/or HO^- adsorbed species into·OH radicals according to Reactions (56) and (57), which will subsequently oxidize the organic compounds. When organic matter is also adsorbed on the surface of the catalyst, it can be directly oxidized by the transfer of an electron from the catalyst [160]. Certainly, in the presence of redox species adsorbed on the semiconductor particle and under irradiation, oxidation and reduction reactions co-occur on the catalyst's surface. The photogenerated holes give rise to photooxidation reactions, while the electrons in the conduction band give rise to photoreduction reactions:

$$TiO_2 + h\nu \rightarrow TiO_2\ (e^-) + TiO_2(h^+) \quad \text{(Reaction 54)}$$

$$TiO_2\ (e^-) + O_2 \rightarrow TiO_2 + O_2^-\cdot \quad \text{(Reaction 55)}$$

$$TiO_2(h^+) + H_2O \rightarrow TiO_2 + OH + H^+ \quad \text{(Reaction 56)}$$

$$TiO_2(h^+) + HO^- \rightarrow TiO_2 + OH \quad \text{(Reaction 57)}$$

The main advantages of this AOP are that it can be operated under pressure and at room temperature, the possibility of using sunlight for the irradiation of the catalyst, and the low cost and reusability of the catalyst. Also, this system is capable of achieving the complete mineralization of many compounds. However, it has significant disadvantages, such as the difficulties of attaining uniform radiation over the entire surface of the catalyst on a larger scale or the need for a subsequent separation treatment to recover the catalyst in suspension, which makes the process more expensive.

5.1. *The TiO_2/UV System*

The process of photocatalytic oxidation, that is, the simultaneous application of air or oxygen, UV radiation, and a semiconductor (mainly TiO_2), is a widely known process that is applied in the removal of numerous compounds but that is not fully developed on a large scale, mostly by the need to separate the photocatalyst, as indicated above. If used in powder (particle size in the range of tens of nm), TiO_2 exhibits high effectiveness. It is also required that the incident radiation on the surface of the photocatalyst has a minimum of energy so that the electrons of the valence band of the semiconductor can be promoted to the conduction band and the generation of hole–electron pairs may take place. In the particular case of titanium dioxide, radiation in the near-ultraviolet is required. This is so because, unfortunately, TiO_2 can absorb no more than 5% of the solar spectrum (i.e., the near UV light with $\lambda < 380$ nm) due to its relatively large bandgap (3.2 eV) [161]. As a consequence, the enhancement of the catalytic activity of TiO_2 within the visible zone of the solar spectrum has received a great deal of attention by the scientific community in recent years. Different strategies have been followed to improve the photocatalytic properties of TiO_2 under visible irradiation, such as surface modification with organic molecules [162] or nanoparticles [163,164] or doping with metal [163,165–167] and non-metal [165,168,169] ions, among others.

From the results summarized in Table 13, it may be concluded that the UV-TiO_2 system is quite useful for the removal of antibiotics from water. However, authors agree that it is not the choice method to be applied for effluents showing high concentrations of pollutants. Furthermore, when large catalyst doses are used, the efficiency of the process decreases. Nevertheless, perhaps the main inconvenience that hinders the applicability of this method is the difficulty of separating and recycling an expensive photocatalyst as TiO_2.

Table 13. Removal efficiency of antibiotics in waters by TiO_2/UV advanced oxidation processes.

Antibiotic	Matrix	Operation Conditions	Maximum Removal Efficiency	Remarks	Reference
Ciprofloxacin (CPR)	Ultrapure water	Graphitized mesoporous carbon (GMC)-TiO_2 nanocomposite used as a catalyst (0.35 g/L) Low-pressure UV lamp (λ = 254 nm)	100% in 45 min	Total mineralization achieved in 90 min Hydroxylation, cleavage of piperazine ring and decarboxylation are the main degradation pathways	[170]
Cloxacillin (CLX)	Ultrapure water; Synthetic pharmaceutical wastewater	TiO_2 = 2.0 g/L; UV light = 150 W	~100%	TiO_2 photocatalysis exhibits larger degradation and mineralization efficiencies than other systems also tested	[171]
Metronidazole (MTR)	Complex aqueous matrix (contains anions, cations, humic acid, and glucose)	TiO_2 = 1.5 g/L UV light intensity = 6.5 mW cm^{-2}	~88% in max 30 min	The presence of common water matrix components hinders drug degradation (except glucose)	[172]
Norfloxacin (NRF)	Ultrapure water	TiO_2 = 0.3 g/L Low-pressure UV lamp (λ = 254 nm)	~90%	TiO_2 photocatalysis is the second most effective method for the removal of NRF, after photo-Fenton (96%)	[173]
Oxacillin (OXC)	Ultrapure water	TiO_2 = 0.5 g/L High-pressure mercury lamp (λ = 365 nm) UV = 150 W	100% in 45 min	~90% mineralization achieved in 135 min	[156]
Oxacillin (OXC)	Synthetic pharmaceutical effluent	TiO_2 = 0.5 g/L High-pressure mercury lamp (λ = 365 nm) UV = 150 W	100% in 45 min (without additives) 100% in 60 min (with additives)	TiO_2 photocatalytic treatment was the least inhibited by additives	[174]

5.2. The TiO_2/H_2O_2/UV System

It has already been stated that the simultaneous presence of H_2O_2 and UV radiation results in the homolytic photodissociation of the hydrogen peroxide molecule, thus giving rise to two hydroxyl radicals according to Reaction (33), see Section 5.1. Furthermore, the addition of hydrogen peroxide to the TiO_2/UV system produces a considerable increase in the photodegradation rate. This effect may be due to the generation of more hydroxyl radicals by the reaction of H_2O_2 with the $TiO_2(e^-)$ generated when titanium dioxide is excited by radiation of the adequate wavelength (not necessarily UV light, as indicated under the previous section).

The reaction between $TiO_2(e^-)$ and H_2O_2 generates additional ·OH radicals easily available to contribute to oxidation processes:

$$TiO_2(e^-) + H_2O_2 \rightarrow TiO_2 + OH^- + OH \quad \text{(Reaction 58)}$$

As it is the case for the TiO_2/UV system, the introduction of doping elements or surface modifications results in a better performance of the TiO_2/H_2O_2/UV system both in terms of removal and mineralization efficiencies. For instance, Jiang et al. [175] have recently reported on the important role played by ferrihydrite (Fh) in the transference of the photo-generated electrons from TiO_2 to H_2O_2. According to these authors, Fh deposited on the surface of the catalyst enables an improved separation of electron–hole pairs. These electrons are more available to be transferred to H_2O_2, thus enhancing its decomposition, Reaction (58), which results in more efficient degradation of the target antibiotic, cefotaxime. The Fh-TiO_2 catalyst is highly active from the catalytic standpoint, is easy to prepare at relatively low cost, and exhibits good stability. Furthermore, according to the experimental data, catalytic activity continues even after the complete decomposition of H_2O_2 has taken place. Of course, the photo-generated electrons can also be directly transferred to dissolved O_2, water, or H_2O_2 so that more hydroxyl and superoxide radicals are generated according to Reactions (55), (56), and (58), respectively.

Similarly, García-Muñoz et al. [176] have recently prepared a mesoporous Fe_2O_3-TiO_2 catalyst that exhibited norfloxacin (NRF) degradation rates more than 60% greater than non-doped mesoporous titania. The maximum enhancement of the degradation rate occurs for 3 wt% Fe_2O_3-TiO_2 catalyst, yielding 100% degradation and 90% mineralization within 120 min of reaction. According to the authors, in the hybrid photocatalyst iron oxidation-reduction reactions take place in the presence of H_2O_2, thus generating $HO_x\cdot$ radical species that also contribute to the removal of NRF:

$$Fe(II)\text{-}TiO_2 + H_2O_2 \rightarrow Fe(III)\text{-}TiO_2 + HO\cdot + OH^- \quad \text{(Reaction 59)}$$

$$Fe(II)\text{-}TiO_2 + H_2O_2 \rightarrow Fe(III)\text{-}TiO_2 + HOO\cdot + H^+ \quad \text{(Reaction 60)}$$

Furthermore, oxidation of NRF in aqueous solutions occurs under mild reaction conditions, namely 25 °C and pH 7.

6. Sonochemical Advanced Oxidation Processes

Ultrasounds (US) are sound waves that have frequencies higher than those that can be perceived by the human ear (16 kHz) and lower than 1 GHz. US can be classified into different categories according to their frequency and intensity.

US of very high frequency and low intensity does not generate physicochemical modifications to the medium in which it is applied. US is used, for example, in medicine for diagnosis. High-intensity US, meanwhile, may interact with the medium. The interactions can be physical so that US is used to ease emulsification, cleaning up, and degassing. If the interactions are chemical, US can find applications in the synthesis of organic compounds and the degradation of pollutants in the environment, among others. This latter field of use of ultrasound is called sonochemistry.

In environmental applications, sonochemistry involves the application of US fields to an effluent, with a frequency between 20 kHz and 2 MHz. Among the various phenomena that appear in the water when an ultrasonic field propagates, ultrasonic cavitation stands out. Cavitation is defined as the phenomenon of formation, growth, and implosion of microbubbles or cavities within the liquid that takes place in a brief time interval (milliseconds) and releases a large amount of energy [158,159].

The generation of the phenomenon of cavitation depends mainly on the frequency and power of the US field. During this process, temperatures close to 5000 °C inside the bubbles and extremely high pressures (100 MPa) are generated locally and in a very short time, conditions that allow complicated chemical reactions to be carried out [177].

An aqueous solution in which cavitation takes place can be assimilated to an environment full of chemical microreactors (the cavitating bubbles) where at least the sonolysis of water takes place; that is, the homolytic breakdown of the molecule into highly reactive ·OH and H· radicals. The subsequent participation of these radicals, especially the OH radicals, in the oxidation of toxic and dangerous molecules that could be found in solution, makes it possible to suggest, at least theoretically, the feasibility of eliminating this type of pollutants without the need to use additional reagents—that can be dangerous—and under mild conditions of temperature or pressure [178]. Thus, the degradation of organic compounds can take place through the action of hydroxyl radicals (an oxidative mechanism) or due to high temperatures (a pyrolytic mechanism) [179].

When US is applied to an aqueous solution of a pollutant, the sonolysis of water may take place [180]:

$$2\,H_2O +))) \rightarrow H_2 + H_2O_2 \quad \text{(Reaction 61)}$$

and the generated H_2O_2 may also undergo sonolysis:

$$H_2O_2 +))) \rightarrow 2\,\cdot OH \quad \text{(Reaction 62)}$$

Recently, Serna et al. [181] have reported the removal of up to 17 pharmaceuticals (nine of them antibiotics) by sonochemical oxidation. The authors propose a double mechanism for the degradation of the pollutants, namely pyrolysis, and interaction between pollutants and hydroxyl radicals. The former mechanism is responsible for the removal of hydrophobic and volatile species that undergo thermal degradation in the liquid–bubble interface. On the contrary, hydrophilic and nonvolatile compounds (including antibiotics) are oxidized by the ·OH radicals in the bulk solution.

However, the degradation efficiency and rate for most of the pollutants are far from being entirely satisfactory. The total mineralization of contaminants by the application of US alone is extremely difficult to achieve, particularly for refractory pollutants. Furthermore, the use of US in AOPs is inefficient from the energy consumption standpoint.

Nevertheless, sonication has successfully been applied as an auxiliary treatment in conjunction with other widely used AOPs and, particularly, with the Fenton reagent and some other derived from it.

For instance, in the so-called sono-Fenton process (Fe^{2+}/H_2O_2/US), the following reactions are proposed [182]. The process initiates with two reactions that are coincident with the first stages of the Fenton process:

$$Fe^{2+} + H_2O_2 \rightarrow Fe^{3+} + OH + OH^- \quad \text{(Reaction 39)}$$

$$H_2O_2 + Fe^{3+} \rightarrow Fe(OOH)^{2+} + H^+ \quad \text{(Reaction 47)}$$

Next, the $Fe(OOH)^{2+}$ decomposes sonochemically giving rise to hydroperoxyl radicals:

$$Fe(OOH)^{2+} +))) \rightarrow Fe^{2+} + HOO \quad \text{(Reaction 63)}$$

Finally, additional sonochemically generated hydroxyl radicals become available for acting as oxidizing agents:

$$Fe^{2+} + HOO\cdot \rightarrow Fe^{2+} + H^+ + O_2 \quad \text{(Reaction 64)}$$

$$Fe^{2+} + H_2O_2 \rightarrow Fe^{3+} + OH^- + HO \quad \text{(Reaction 39)}$$

According to Serna et al. [181] the removal of the pollutants in the effluent is more effective in the sono-Fenton process when compared with sonochemical oxidation alone, due to the generation of extra ·OH through reactions between Fe(II) and the sonogenerated hydrogen peroxide.

A similar situation has been described for the nZVI-based hetero-Fenton process, where nanoscaled Fe^0 particles substitute Fe(II) in the initial stages of the process [182] as well as in the sono-Fenton-like process, where Fe(III) salts are used instead of Fe(II) [183]. This suggests that the use of a catalyst may improve energy consumption due to the occurrence of synergistic effects.

Zhou et al. [184] reported on the use of a goethite catalyst in the heterogeneous sonophotolytic Fenton-like (SP-FL) treatment of antibiotic sulfamethazine. The authors integrated the in situ H_2O_2 generation under UV illumination with a heterogeneous Fenton-like process and suggested that the synergistic role of US in the SP-FL system could be most ascribed to its promotional effect in Fenton-like reaction. Efficient Fe(II) species regeneration, improved mineralization degree, and successful wastewater detoxification were achieved.

Synergy appears to be more remarkable as the particle size of the catalyst decreases. Hence, the use of nanostructured catalysts is receiving a great deal of attention from the scientific community at present. Ghoreishian et al. [185] recently used flowerlike rGO/$CdWO_4$ solar-light-responsive photocatalysts for the US-assisted heterogeneous degradation of tetracycline. Excellent photoelectrochemical behavior, superior sonophotocatalytic activity, and good mineralization efficiency were achieved under optimal conditions. Tetracycline was removed entirely at a pH of 5.7, an initial antibiotic concentration of 13.54 mg L^{-1}, a treatment time of 60 min, and a catalyst dosage of 0.216 g L^{-1}. Furthermore, rGO/$CdWO_4$ exhibited a sonophotocatalytic efficiency that was 1.5 and 3 times higher than commercial nano–ZnO and nano–TiO_2, respectively. Tetracycline was also chosen as the target pollutant by Vinesh et al. [186], which used a reduced graphene oxide (rGO)-supported electron-deficient B-doped TiO_2 (Au/B-TiO_2/rGO) nanocomposite. A considerable synergistic effect of ~1.3 was observed when the reaction was performed in the presence of US and photocatalysis. The total degradation of the antibiotic was also confirmed by TOC analysis. The enhanced sonophotocatalytic activity was mainly attributed to the generation of more reactive species by the combination of US and photocatalysis.

Also, Abazari et al. [187] applied Ni–Ti layered double hydroxide@graphitic carbon nitride nanosheet for photocatalytic and sonophotocatalytic removal of amoxicillin from aqueous solution. The authors related the enhancement in the sonophotocatalytic activity of the nanocomposites to their higher specific surface areas, the intimacy of the contact interfaces of their components, the synergistic effect between these components, and the restriction of electron–hole recombination. The optimum sonophotocatalysis conditions were 500 W light intensity, 9 s on/1 s off US pulse mode, and 1.25 g/L of g-C_3N_4@Ni–Ti LDH catalyst. Under these conditions, 99.5% removal within 75 min was attained.

7. Electro-Oxidative Advanced Oxidation Processes.

The application of electric current (from 2 to 20 A) between two electrodes in water produces the generation of hydroxyl ·OH radicals coupled with the production of hydrogen peroxide in the reaction medium.

$$H_2O \rightarrow OH + H^+ + e^- \quad \text{(Reaction 65)}$$

$$O_2 + 2\,H^+ + 2\,e^- \rightarrow H_2O_2 \quad \text{(Reaction 66)}$$

The global reaction is:

$$2\,H_2O + O_2 \rightarrow H_2O_2 + 2\,OH \quad \text{(Reaction 67)}$$

Hence, the OH radicals can be regarded as the product of the anodic oxidation of water and are readily available to oxidize the organic matter in aqueous solution. At the same time, hydrogen peroxide is generated by cathodic reduction of oxygen [188].

Electro-oxidation, also called anodic oxidation or electrochemical incineration, is one of the most popular electrochemical advanced oxidation processes for the elimination of organic contaminants contained in wastewater [28,189].

This procedure involves the oxidation of the contaminants in an electrolytic cell through the following routes:

(a) Direct electron transfer to the anode.
(b) Indirect or mediated oxidation with oxidizing species formed from the electrolysis of water at the anode, by physisorbed OH radicals or by the chemisorbed "active oxygen."

The existence of these species allows two different approaches to be proposed [190]:

(1) Electrochemical conversion, where the refractory organic compounds are selectively transformed into biodegradable compounds, such as carboxylic acids, by the chemisorbed "active oxygen."
(2) Electrochemical combustion, where the physisorbed OH radicals mineralize the organic compounds.

Anodic oxidation can achieve the oxidation of water pollutants, either by direct contact or by oxidative processes that take place on the anodic surface of the electrochemical cell. Hence, the oxidation process does not necessarily have to occur in the anode, but it is initiated on its surface. As a consequence, this treatment combines two main types of processes [191]:

a. Heterogeneous oxidation of contaminants on the surface of the anode. This is a complex process that consists of a series of simple steps: The transport of pollutants to the surface of the electrode, the adsorption of the contaminant on the electrode's surface, the direct electrochemical reaction by electron transfer, the desorption of products, and the transport of such products to the dissolution.
b. Homogeneous oxidation of contaminants by oxidants produced on the surface of the anode. These oxidizing species can be produced by the heterogeneous anodic oxidation of the water or from ions contained in the water acting in the dissolution of the electrolytic cell. The most important oxidant is the hydroxyl radical, which can be generated by the oxidation of water, Reaction (65), or by oxidation of the hydroxyl ion, Reaction (68):

$$H_2O \rightarrow \cdot OH_{ads} + H^+ + e^- \quad \text{(Reaction 65)}$$

$$OH^- \rightarrow OH_{ads} + e^- \quad \text{(Reaction 68)}$$

The generation of this OH radical is the main argument for considering anodic oxidation as an AOP. Due to the high oxidation capacity of hydroxyl radicals, these promote the formation of many other oxidizing species (persulfates, peroxophosphates, ferrates, etc.) from different species contained in real water matrices [192]. It has been shown that the presence of these species has a significant effect on the increase in degradation efficiency [193]. The synergistic effects of all these mechanisms can explain the high efficiencies obtained in the elimination of pollutants and the high mineralization achieved with this technology in comparison with other AOPs [194].

Some materials lead to powerful oxidation of the pollutant, yielding CO_2 and H_2O as the major final products and a small number of intermediates, while other materials produce less oxidation and generate a large number of oxidation byproducts. Comninellis [195] proposed an integral model for the destruction of organic compounds in an acidic medium that assumes the existence of "non-active" and "active" anodes. According to this model, "active" anodes, which promote the electrochemical reaction of oxygen evolution, favor the electrochemical conversion of organic matter, while the "non-active", which are less electrocatalytic for the evolution of oxygen, requiring higher anodic overpotentials, favor the electrochemical combustion of organic matter. In both types of anodes, denoted as M, the water oxidizes giving rise to the formation of physisorbed hydroxyl radicals (M (OH)):

$$M + H_2O \rightarrow M(\cdot OH)_{physisorbed} + H^+ + e^- \quad \text{(Reaction 69)}$$

In the case of "active" anodes, this radical interacts strongly with the surface, transforming into chemisorbed "active oxygen" or superoxide MO:

$$M(OH)_{physisorbed} \rightarrow MO_{chemisorbed} + H^+ + e^- \quad \text{(Reaction 70)}$$

The MO/M pair is a mediator in the electrochemical conversion of organic compounds (R):

$$MO + R \rightarrow M + RO \quad \text{(Reaction 71)}$$

On the contrary, the surface of the "non-active" anodes interacts weakly with the OH species, so that these radicals react directly with the organic products until total mineralization is at least ideally achieved. These radicals are physisorbed on electroactive electrode sites and do not undergo modification during the electron transfer reaction.

Graphite electrodes, with sp^2 carbon, metal (Pt, Ti/Pt), metal oxide (IrO_2, RuO_2), and mixed metal oxides electrodes, are considered as "active" anodes and behave as low-efficiency electrodes for the oxidation of organic compounds, generating a large number of intermediate byproducts. Most aromatic compounds treated with these anodes degrade slowly due to the generation of carboxylic acids that are difficult to oxidize [196]. Small mineralization efficiencies are obtained, and in some cases, polymeric species are generated, thus hindering the perspectives of application of these materials as anodes for electrooxidation processes.

Some metallic oxides and mixed metal oxides (those containing PbO_2 and/or SnO_2) and conductive diamond electrodes, in particular, the boron-doped diamond (BDD) electrodes are considered "non-active" anodes and behave as highly efficient electrodes for the oxidation of organic compounds. These anodes promote the mineralization of pollutants, whose extent is limited only by the mass transfer, and in general, virtually total mineralization of the contaminant is achieved. The surface material of BDD electrodes represents a promising technology for the electroanalysis of different biologically relevant active compounds [197–201].

In practice, all anodes exhibit a mixed behavior, since both mechanisms take place simultaneously. The "non-active" anodes may have defects in their surface or partially oxidized sites, while in the case of the "active" electrodes, the formation of physisorbed radicals at very positive potentials cannot be excluded, even if the surface is highly reactive. It should be mentioned that, as a general rule, the less positive the potential at which the evolution of oxygen occurs, the higher the participation of the anode surface in the reaction.

The generalized reaction pathway for the oxidation of an organic compound, R, and the simultaneous electrochemical formation of oxygen (over an anode, M), for the two types of anodes proposed, includes the following stages [202]:

(1) Generation and adsorption of the·OH radical.
(2) Oxygen production by electrochemical oxidation of adsorbed·OH radicals.
(3) Formation of a site with a higher oxidation state by electrochemical oxidation of the radical OH.
(4) Production of oxygen by chemical decomposition of the site with a higher oxidation state.
(5) Combustion of the organic compound, R, by physisorbed·OH radicals.
(6) Chemical oxidation of the organic compound at a site with a higher oxidation state.

The BDD anode is the most potent known "non-active" electrode [191]. It is considered as the most suitable anode for the treatment of organic compounds by anodic oxidation. Furthermore, the BDD electrodes have high anodic stability and a broad working potential range [203,204]. The use of BDD has considerably increased the interest in the application of this method in the treatment of waters since excellent mineralization efficiencies are obtained.

It should be noted that this anode exhibits excellent chemical and electrochemical stability, an inert surface with low adsorption properties, long life, and a wide range of potential for water discharge [205, 206], and therefore, it turns out to be a promising electrode for the decontamination treatment.

The main limitation of technology based on BDD electrodes is its high price, which hinders its use at the industrial scale. It has been shown that many biorecalcitant compounds including phenols, chlorophenols, nitrophenols, pesticides, synthetic dyes, pharmaceuticals, and industrial leachates can be completely mineralized with high current efficiency, even close to 100%, using a BDD anode [207–213].

The composition of the supporting electrolyte or the different ions present in real industrial effluents can vary the effectiveness of the electrooxidation process. During the electrolysis with BDD anodes, simultaneously with the generation of active oxygen species that give rise to the generation of ·OH radicals, numerous reactions take place that can also lead to other oxidants depending on the ions present in the treated volume. In sulfate medium, oxidation of sulfate ions to peroxodisulfate can occur [209,214]:

$$2\,SO_4^{2-} \rightarrow S_2O_8^{2-} + 2\,e^- \quad \text{(Reaction 72)}$$

A very particular behavior is found when the solution contains chloride ions since the electrogeneration of active chlorine in the form of chlorine gas, hypochlorous acid, or hypochlorite ion occurs through Reactions (73)–(75). Under these conditions, organic matter can be competitively attacked by the ·OH radicals produced on the surface of the anode and the active chlorine produced and diffused into the solution. As a counterpoint, the formation of organochlorine intermediates can occur, which can be even more stubborn and more toxic than the primary pollutants.

$$2\,Cl^- \rightarrow Cl_{2(aq)} + 2\,e^- \quad \text{(Reaction 73)}$$

$$Cl_{2(aq)} + H_2O \leftrightarrow HClO + Cl^- + H^+ \quad \text{(Reaction 74)}$$

$$HClO \leftrightarrow ClO^- + H^+ \quad \text{(Reaction 75)}$$

When a "non-active" anode such as BDD is used, hypochlorite ions are also generated from the oxidation of chloride ions by the ·OH radicals adsorbed on the BDD anode according to Reaction (76). The resulting ion can be oxidized consecutively to chlorite, chlorate, and perchlorate according to Reactions (76)–(79) [215,216]:

$$Cl^- + OH \rightarrow ClO^- + H^+ + e^- \quad \text{(Reaction 76)}$$

$$ClO^- + \cdot OH \rightarrow ClO_2^- + H^+ + e^- \quad \text{(Reaction 77)}$$

$$ClO_2^- + OH \rightarrow ClO_3^- + H^+ + e^- \quad \text{(Reaction 78)}$$

$$ClO_3^- + OH \rightarrow ClO_4^- + H^+ + e^- \quad \text{(Reaction 79)}$$

It is worth mentioning that one of the most advanced large-scale applications in the field of electrochemical AOPs is the electrochemical disinfection of pool and spa water using automated equipment with BDD anodes. In this field, specialized products such as Oxineo™ and Sysneo™ have been developed for private and public facilities. Compared to other methods of disinfection, these systems have several advantages, since the chlorine smell typical of saline chlorination disappears, there is no accumulation of chemicals in the pool water—with the consequent reduction of allergic reactions—and there is no need to use anti-algae chemicals.

8. Prospects and Challenges of AOPs

From all the exposed in this review, AOPs arise as beneficial technologies for the removal of pollutants in general and antibiotics, in particular, that may be found in wastewaters. However, the applicability of these processes at the industrial scale is relatively limited at present.

Very recently, Rodríguez-Chueca et al. [217] have reported that AOPs, in general, improved the efficiency of UV-C on the removal of antibiotics. When applying the harshest operating conditions (0.5 mM dosage of oxidants and 7s of UV-C contact time), three antibiotics (ciprofloxacin, metronidazole, and sulfamethoxazole) were similarly removed using peroxymonosulfate and H_2O_2 as oxidants, four antibiotics (azithromycin, clindamycin, ofloxacin, and trimethoprim) were removed more efficiently using peroxymonosulfate, and three (clarithromycin, sulfadiazine, and sulfapyridine) using H_2O_2. The addition of Fe(II) only improved the degradation of four antibiotics (sulfamethoxazole, sulfadiazine, sulfapyridine, and metronidazole) compared to the photolytic process.

Chowdhury et al. [218] have studied the direct UV photolysis of different pharmaceutical compounds, including tetracycline antibiotics (chlortetracycline), sulfa drugs (sulfamethoxazole, sulfathiazole, and sulfisoxazole), and fluoroquinolones (ciprofloxacin) at full-scale. Except for ciprofloxacin, the remaining drugs showed considerable pH-dependent photolysis in the pH range under study (namely, 5–8). High removal by UV photolysis was attained, although a much higher UV fluence than that used for water disinfection was required. This latter finding suggests that additional treatments are necessary for water recycling.

Östman et al. [67] analyzed the effect of full-scale ozonation on the removal of up to eight antibiotics (namely, ciprofloxacin, norfloxacin, ofloxacin, pentamidine, clarithromycin, erythromycin, metronidazole, and trimethoprim) in Sweden's first sewage treatment plant with full-scale ozonation. Ozonation proved to be effective in most cases. Total (100%) removal of ciprofloxacin and metronidazole was reported, whereas 94% was achieved for erythromycin (versus 79% in the conventional treatment). Trimethoprim was poorly removed (41%) in the traditional treatment process, but 94% removal efficiency was reached by ozonation.

Rame et al. [219] also indicate that FLASH technology that uses physical and biological pre-treatment, followed by an advanced oxidation process based on catalytic ozonation and followed by GAC and PAC filtration, has demonstrated good removal efficiency of macro-pollutants present in hospital wastewaters, including antibiotics.

Some years before, Sui et al. [220] reported that the sequential UV and ozonation process in a full-scale WWTP in Beijing was able to remove up to 13 pharmaceuticals and personal care products

(PPCPs) including chloramphenicol. According to these authors, notwithstanding the fact that the UV exhibited a limited removal ability for most of the PPCPs under study, the sequential use of UV and ozone made it possible to minimize PPCPs in the final effluent of WWTPs.

To increase the feasibility of AOPs to be used at large scale, several aspects should be improved. With such an aim, further research work should be—and is currently being—performed on different topics such as, for instance, costs of the process, toxicity of effluents and byproducts, (photo)catalysts technology, and reactors design.

Many papers dealing with operational costs of AOPs at bench scale are available. However, articles focusing on operational expenses of full-scale advanced oxidation processes are scarce, and attention has been paid to this issue only in the last few years [46,221–227]. From the literature review, it may be concluded that treatment costs tend to decrease as the pollutant's concentration in water increases. Moreover, the ozonation and Fenton processes appear to be more economically viable and, as a consequence, have been more frequently implemented at the industrial scale. However, authors do not agree on which of these treatments is the most cost-effective one. When integrating UV radiation to other AOPs, it can be stated that removal efficiencies increase, but operational costs are higher, too. Also, it is commonly accepted that photocatalysis (mainly using TiO_2) is a relatively expensive AOP but also quite useful for the removal of pollutants.

In many cases, AOPs can be successfully applied for the removal of pollutants present in wastewaters and chemically degrade the parent compounds leading to their complete mineralization, thus generating CO_2 and water as the leading products and without generation of any toxicity. However, it is not infrequent that the oxidation leads to different byproducts that may have similar—or even higher—toxicity than the original pollutants themselves. Changes in the molecular structure of the pollutant may give rise to an entirely new kind of chemical toxicity or even to mutagenic [228] or estrogenic activity [229]. For instance, during ozonation, the bromide naturally occurring in wastewater can be oxidized to bromate [230], with increasing yields as ozone dose rises [231]. Bromate is included in the EPA's and EU's lists of potential carcinogens. However, the addition of H_2O_2 largely mitigates the formation of bromate [232]. Similarly, if wastewaters contain large amounts of nitrate, the treatment by UV/H_2O_2 AOP may result in increased mutagenicity of the treated effluent [233]. According to the literature, the incorporation of the nitrate-nitrogen into the organic matrix and the subsequent formation on nitrated/nitrosated compounds could be the reason that mutagenicity levels increase in treated waters. Other authors have reported similar results in terms of increased toxicity after the treatment of wastewaters containing antibiotics [77,234–238].

Among the different alternatives to generate $\cdot OH$ radicals in AOPs, heterogeneous photocatalysis appears as an up-and-coming solution in wastewater treatment. Photocatalysis is a nonselective process that can degrade a wide variety of pollutants. A photocatalyst (i.e., a semiconductor material) and light are the only requirements of this process. However, to the date, most of the results published in the literature focus on the use of UV irradiation, with wavelengths ranging from 320 to 400 nm, mainly because the bandgap of the semiconductor must be high enough to avoid fast recombination of the electron/hole pairs generated when the photoexcitation of the catalyst takes place. Different metal oxide semiconductors have been tested and used as potential photocatalysts for wastewater treatment. According to the literature [239], TiO_2 and ZnO are the most widely used metal oxides in AOPs, even though other metal oxides (such as V_2O_5, WO_3, MoO_3, or some of their derivatives) or even sulfides (CdS, ZnS) could also be used. However, their performance is below that of TiO_2. Furthermore, TiO_2 exhibits essential advantages: It can be used under ambient conditions, is relatively inexpensive, commercially available, non-toxic, and photochemically stable.

An adequate modification of the chemical composition of the catalyst may result in enhanced efficiency and reusability. For instance, noble metals (Au, Pt, Ag, and Pd), transition metals (Cr, Mn, Fe, Co, Ni, Cu, and Zn) or lanthanides (La, Nd, Sm, Eu, Gd, Yb, or Pr) have been used as dopants in order to improve the photocatalytic activity of TiO_2. The main goals of the chemical modification of TiO_2 by using these dopants are: (i) To reduce the bandgap of the photocatalyst so that it is compatible with solar light (and not only with UV) to a solar light-compatible level; (ii) to maximize electron–hole generation; and (iii) to increase the adsorption ability of TiO_2 towards organic pollutants by increasing its specific surface area [240–242].

Recently, a new group of materials, commonly known as two-dimensional (2D) nanomaterials, has shown promising properties to be used as novel photocatalysts and are receiving increasing attention from the scientific community [243]. These nanomaterials exhibit unique optical, electronic, and physicochemical properties, but their eventual applicability may be limited due to high production costs. 2D nanomaterials include graphene, graphitic carbon nitride (g-C_3N_4), 2D metal oxides and metallates, metal oxyhalides, and transition metal dichalcogenides. These materials can be applied alone as highly efficient photocatalysts. However, some of them exhibit better performance if they are combined with other "traditional" photocatalysts (e.g., TiO_2), which, in turn, make them more affordable. For instance, it has been demonstrated that the use of graphene and TiO_2 in the form of a heterojunction provides much higher photocatalytic efficiency than pure TiO_2 [244].

Finally, another critical aspect to be taken into consideration is the proper design of the reactor. Fluidized bed reactor (FBR) has been successfully used in the full-scale application of AOPs in treatment plants [245] and is perhaps the most versatile option for this kind of processes. FBR has proven to be more effective in wastewater treatment compared to other conventional reactors such as, for instance, fixed-bed column and activated sludge. If FBR technology is combined with AOPs sludge production—one of the main disadvantages of the Fenton process—can be minimized [246], the reusability of the catalyst is increased [245], and the overall performance of the process is improved. FBRs exhibit remarkable advantages such as low operating cost, high resistance, uniformity of mixing, and high mass transfer rates. This latter is particularly important in the implementation of full-scale O_3-based AOPS since one of the most significant drawbacks of ozone technologies is the difficulty of achieving an adequate mass transfer of gaseous ozone into the bulk solution. The most commonly investigated FBR-integrated AOPs are fluidized bed Fenton and fluidized bed photocatalysis. In the case of FBR heterogeneous Fenton process, high degradation efficiency has been achieved and the sludge generation has been reduced simultaneously [247]. On the other hand, in addition to its excellent mixing and mass transfer ability, in photocatalytic processes, the use of FBR can also enhance light penetration and exposure of the interior of the reaction matrix, thus improving the overall yield of the process in terms of pollutant(s) degradation.

9. Conclusions and Final Recommendations

Antibiotics are almost ubiquitous pollutants that have been found in different kinds of surface waters, wastewaters, and WWTP and hospital effluents. Furthermore, the presence of antibiotics in water may cause harmful effects to human beings as well as promote the spread of resistant bacterial strains. After several decades of research on AOPs, these technologies have proven their efficiency for the removal of a wide variety of pollutants in general and antibiotics in particular. However, most of the literature published to the date is devoted to bench- or pilot-scale studies. The implementation of AOPs at the full-scale is still quite limited. Probably, the main difficulty for the development of these processes at an industrially operative scale is the high operational cost of AOPs, especially if compared with the conventional methods that are commonly applied nowadays. Thus, if the overall cost per unit mass of pollutant that is removed from water or unit volume of water, wastewater, or effluent that is treated is lowered, the industrial implementation of these technologies will become much more attractive for companies and/or public administrations. Some suggestions on the way to achieve this goal are as follows:

(i) To avoid unnecessary expenses in terms of time, facilities, and reagents, AOPs should be integrated with other treatments and only with specific and clearly defined goals (i.e., the removal of recalcitrant (micro)pollutants, the polishing of previously treated effluents, etc.). Furthermore, synergistic effects between processes should be studied at least at the pilot-scale.

(ii) Energy costs must also be reduced. In this connection, the search for novel, affordable photocatalysts that may use a broader part of the light spectrum instead of only UV is a priority. Furthermore, the application of renewable energy sources in the treatment plants should also be investigated.
(iii) The generation of wastes (e.g., sludge in the Fenton process and/or exhausted or poisoned catalysts in photocatalyzed AOPs), should be minimized and possible alternatives for the valorization of such wastes should be explored.

Author Contributions: Conceptualization, E.M.C.-C.; bibliographical search, E.M.C.-C., M.F.A.-F. and C.F.-G.; formal analysis of Photolysis and Ozone-based AOPs, E.M.C.-C. and M.F.A.-F.; formal analysis of Hydrogen peroxide-based AOPs, M.F.A.-F. and C.F.-G.; formal analysis of Heterogeneous photocatalysis, E.M.C.-C. and C.F.-G.; formal analysis of Sonochemical and Electro-oxidative AOPs, E.M.C.-C. and M.F.A.-F.; formal analysis of Prospects and challenges of AOPs, E.M.C.-C., M.F.A.-F. and C.F.-G.; writing—original draft preparation, E.M.C.-C., M.F.A.-F. and C.F.-G.; writing—review and editing, E.M.C.-C; supervision, E.M.C.-C. All authors have read and agreed to the published version of the manuscript.

Funding: This research received no external funding.

Conflicts of Interest: The authors declare no conflict of interest.

References

1. NESCO and World Health Organization. *Progress on Drinking Water and Sanitation: 2012 Update*; UNICEF, Division of Communication: New York, NY, USA, 2012; ISBN 9789280646320.
2. UNESCO. *The 1st UN World Water Development Summary Report*; UNESCO: London, UK, 2003; Volume 36.
3. United Nations. *Water A Shared Responsibility The United Nations World Water Development Report 2 Berghahn Books*; UNICEF, Division of Communication: San Francisco, CA, USA, 2006; ISBN 9231040065.
4. Brillas, E.; Sirés, I.; Oturan, M.A. Electro-fenton process and related electrochemical technologies based on fenton's reaction chemistry. *Chem. Rev.* **2009**, *109*, 6570–6631. [CrossRef] [PubMed]
5. Relyea, R.A. New effects of Roundup on amphibians: Predators reduce herbicide mortality; Herbicides induce antipredator morphology. *Ecol. Appl.* **2012**, *22*, 634–647. [CrossRef] [PubMed]
6. Petrović, M.; Gonzalez, S.; Barceló, D. Analysis and removal of emerging contaminants in wastewater and drinking water. *TrAC Trends Anal. Chem.* **2003**, *22*, 685–696. [CrossRef]
7. Farré, M.L.; Pérez, S.; Kantiani, L.; Barceló, D. Fate and toxicity of emerging pollutants, their metabolites and transformation products in the aquatic environment. *TrAC Trends Anal. Chem.* **2008**, *27*, 991–1007. [CrossRef]
8. Gagné, F.; Blaise, C.; André, C. Occurrence of pharmaceutical products in a municipal effluent and toxicity to rainbow trout (Oncorhynchus mykiss) hepatocytes. *Ecotoxicol. Environ. Saf.* **2006**, *64*, 329–336. [CrossRef]
9. Halling-Sørensen, B.; Nors Nielsen, S.; Lanzky, P.F.; Ingerslev, F.; Holten Lützhøft, H.C.; Jørgensen, S.E. Occurrence, fate and effects of pharmaceutical substances in the environment—A review. *Chemosphere* **1998**, *36*, 357–393. [CrossRef]
10. Huber, M.M.; Göbel, A.; Joss, A.; Hermann, N.; Löffler, D.; McArdell, C.S.; Ried, A.; Siegrist, H.; Ternes, T.A.; Von Gunten, U. Oxidation of pharmaceuticals during ozonation of municipal wastewater effluents: A pilot study. *Environ. Sci. Technol.* **2005**, *39*, 4290–4299. [CrossRef]
11. Santos, L.H.M.L.M.; Araújo, A.N.; Fachini, A.; Pena, A.; Delerue-Matos, C.; Montenegro, M.C.B.S.M. Ecotoxicological aspects related to the presence of pharmaceuticals in the aquatic environment. *J. Hazard. Mater.* **2010**, *175*, 45–95. [CrossRef]
12. Joss, A.; Keller, E.; Alder, A.C.; Göbel, A.; McArdell, C.S.; Ternes, T.; Siegrist, H. Removal of pharmaceuticals and fragrances in biological wastewater treatment. *Water Res.* **2005**, *39*, 3139–3152. [CrossRef]
13. Joss, A.; Zabczynski, S.; Göbel, A.; Hoffmann, B.; Löffler, D.; McArdell, C.S.; Ternes, T.A.; Thomsen, A.; Siegrist, H. Biological degradation of pharmaceuticals in municipal wastewater treatment: Proposing a classification scheme. *Water Res.* **2006**, *40*, 1686–1696. [CrossRef]
14. Lee, Y.; Von Gunten, U. Oxidative transformation of micropollutants during municipal wastewater treatment: Comparison of kinetic aspects of selective (chlorine, chlorine dioxide, ferrateVI, and ozone) and non-selective oxidants (hydroxyl radical). *Water Res.* **2010**, *44*, 555–566. [CrossRef] [PubMed]

15. Fewson, C.A. Biodegradation of xenobiotic and other persistent compounds: The causes of recalcitrance. *Trends Biotechnol.* **1988**, *6*, 148–153. [CrossRef]
16. López, R.; Menéndez, M.I.; Díaz, N.; Suárez, D.; Campomanes, P.; Ardura, D.; Sordo, T.L. Theoretical studies on the ring opening of β-lactams: Processes in solution and in enzymatic media. *Curr. Org. Chem.* **2006**, *10*, 805–821. [CrossRef]
17. Robinson, T.; McMullan, G.; Marchant, R.; Nigam, P. Remediation of dyes in textile effluent: A critical review on current treatment technologies with a proposed alternative. *Bioresour. Technol.* **2001**, *77*, 247–255. [CrossRef]
18. Forgacs, E.; Cserháti, T.; Oros, G. Removal of synthetic dyes from wastewaters: A review. *Environ. Int.* **2004**, *30*, 953–971. [CrossRef]
19. Glaze, W.H.; Kang, J.W.; Chapin, D.H. The chemistry of water treatment processes involving ozone, hydrogen peroxide and ultraviolet radiation. *Ozone Sci. Eng.* **1987**, *9*, 335–352. [CrossRef]
20. Ayoub, K.; van Hullebusch, E.D.; Cassir, M.; Bermond, A. Application of advanced oxidation processes for TNT removal: A review. *J. Hazard. Mater.* **2010**, *178*, 10–28. [CrossRef]
21. Legrini, O.; Oliveros, E.; Braun, A.M. Photochemical Processes for Water Treatment. *Chem. Rev.* **1993**, *93*, 671–698. [CrossRef]
22. Klavarioti, M.; Mantzavinos, D.; Kassinos, D. Removal of residual pharmaceuticals from aqueous systems by advanced oxidation processes. *Environ. Int.* **2009**, *35*, 402–417. [CrossRef]
23. Malato, S.; Fernández-Ibáñez, P.; Maldonado, M.I.; Blanco, J.; Gernjak, W. Decontamination and disinfection of water by solar photocatalysis: Recent overview and trends. *Catal. Today* **2009**, *147*, 1–59. [CrossRef]
24. Haag, W.R.; David Yao, C.C. Rate Constants for Reaction of Hydroxyl Radicals with Several Drinking Water Contaminants. *Environ. Sci. Technol.* **1992**, *26*, 1005–1013. [CrossRef]
25. Garcia-Segura, S.; El-Ghenymy, A.; Centellas, F.; Rodríguez, R.M.; Arias, C.; Garrido, J.A.; Cabot, P.L.; Brillas, E. Comparative degradation of the diazo dye Direct Yellow 4 by electro-Fenton, photoelectro-Fenton and photo-assisted electro-Fenton. *J. Electroanal. Chem.* **2012**, *681*, 36–43. [CrossRef]
26. Guinea, E.; Arias, C.; Cabot, P.L.; Garrido, J.A.; Rodríguez, R.M.; Centellas, F.; Brillas, E. Mineralization of salicylic acid in acidic aqueous medium by electrochemical advanced oxidation processes using platinum and boron-doped diamond as anode and cathodically generated hydrogen peroxide. *Water Res.* **2008**, *42*, 499–511. [CrossRef] [PubMed]
27. Pera-Titus, M.; García-Molina, V.; Baños, M.A.; Giménez, J.; Esplugas, S. Degradation of chlorophenols by means of advanced oxidation processes: A general review. *Appl. Catal. B Environ.* **2004**, *47*, 219–256. [CrossRef]
28. Martínez-Huitle, C.A.; Ferro, S. Electrochemical oxidation of organic pollutants for the wastewater treatment: Direct and indirect processes. *Chem. Soc. Rev.* **2006**, *35*, 1324–1340. [CrossRef]
29. Serra, A.; Domènech, X.; Arias, C.; Brillas, E.; Peral, J. Oxidation of α-methylphenylglycine under Fenton and electro-Fenton conditions in the dark and in the presence of solar light. *Appl. Catal. B Environ.* **2009**, *89*, 12–21. [CrossRef]
30. Feng, L.; Van Hullebusch, E.D.; Rodrigo, M.A.; Esposito, G.; Oturan, M.A. Removal of residual anti-inflammatory and analgesic pharmaceuticals from aqueous systems by electrochemical advanced oxidation processes. A review. *Chem. Eng. J.* **2013**, *228*, 944–964. [CrossRef]
31. Sirés, I.; Brillas, E.; Oturan, M.A.; Rodrigo, M.A.; Panizza, M. Electrochemical advanced oxidation processes: Today and tomorrow. A review. *Environ. Sci. Pollut. Res.* **2014**, *21*, 8336–8367. [CrossRef]
32. Andreozzi, R.; Caprio, V.; Insola, A.; Marotta, R. Advanced oxidation processes (AOP) for water purification and recovery. *Catal. Today* **1999**, *53*, 51–59. [CrossRef]
33. Bokare, A.D.; Choi, W. Review of iron-free Fenton-like systems for activating H_2O_2 in advanced oxidation processes. *J. Hazard. Mater.* **2014**, *275*, 121–135. [CrossRef]
34. Ince, N.H.; Apikyan, I.G. Combination of activated carbon adsorption with light-enhanced chemical oxidation via hydrogen peroxide. *Water Res.* **2000**, *34*, 4169–4176. [CrossRef]
35. Anjali, R.; Shanthakumar, S. Insights on the current status of occurrence and removal of antibiotics in wastewater by advanced oxidation processes. *J. Environ. Manag.* **2019**, *246*, 51–62. [CrossRef] [PubMed]
36. Bartolomeu, M.; Neves, M.G.P.M.S.; Faustino, M.A.F.; Almeida, A. Wastewater chemical contaminants: Remediation by advanced oxidation processes. *Photochem. Photobiol. Sci.* **2018**, *17*, 1573–1598. [CrossRef] [PubMed]

37. Bousiakou, L.; Kazi, M.; Lianos, P. Wastewater treatment technologies in the degradation of hormones and pharmaceuticals with focus on TiO_2 technologies. *Pharmakeftiki* **2013**, *25*, 37–48.
38. Daghrir, P.; Drogui, R. Tetracycline antibiotics in the environment: A review. *Environ. Chem. Lett.* **2013**, *11*, 209–227. [CrossRef]
39. Rivera-Utrilla, J.; Sánchez-Polo, M.; Ferro-García, M.Á.; Prados-Joya, G. Pharmaceuticals as emerging contaminants and their removal from water: A review. *Chemosphere* **2013**, *93*, 1268–1287. [CrossRef]
40. Oller, I.; Malato, S.; Sánchez-pérez, J.A. Combination of Advanced Oxidation Processes and biological treatments for wastewater decontamination—A review. *Sci. Total Environ.* **2011**, *409*, 4141–4166. [CrossRef]
41. Homem, V.; Santos, L. Degradation and removal methods of antibiotics from aqueous matrices—A review. *J. Environ. Manag.* **2011**, *92*, 2304–2347. [CrossRef]
42. Yargeau, V.; Leclair, C.; Yargeau, V.; Leclair, C. Impact of Operating Conditions on Decomposition of Antibiotics During Ozonation: A Review Impact of Operating Conditions on Decomposition of Antibiotics During Ozonation: A Review. *Ozone Sci. Eng.* **2008**, *30*, 175–188. [CrossRef]
43. Esplugas, S.; Bila, D.M.; Gustavo, L.; Krause, T. Ozonation and advanced oxidation technologies to remove endocrine disrupting chemicals (EDCs) and pharmaceuticals and personal care products (PPCPs) in water effluents. *J. Hazard. Mater.* **2007**, *149*, 631–642. [CrossRef]
44. Ikehata, K.; Naghashkar, N.J.; El-din, M.G.; Ikehata, K.; Naghashkar, N.J.; El-din, M.G. Degradation of Aqueous Pharmaceuticals by Ozonation and Advanced Oxidation Processes: A Review. *Ozone Sci. Eng.* **2007**, *28*, 353–414. [CrossRef]
45. Jain, B.; Kumar, A.; Hyunook, S.; Eric, K.; Virender, L. Treatment of organic pollutants by homogeneous and heterogeneous Fenton reaction processes. *Environ. Chem. Lett.* **2018**, *16*, 947–967. [CrossRef]
46. Lofrano, G.; Pedrazzani, R.; Libralato, G.; Carotenuto, M. Advanced Oxidation Processes for Antibiotics Removal: A Review. *Curr. Org. Chem.* **2017**, *21*, 1–14. [CrossRef]
47. Yan, S.; Song, W. Photo-transformation of pharmaceutically active compounds in the aqueous environment: A review. *Environ. Sci. Process. Impacts* **2014**, *16*, 697–720. [CrossRef]
48. Oppenländer, T. *Photochemical Purification of Water and Air: Advanced Oxidation Processes (AOPs): Principles, Reaction Mechanisms, Reactor Concepts*; Wiley-VCH Verlag GmbH & Co.: Weinheim, Germany, 2002.
49. Timm, A.; Borowska, E.; Majewsky, M.; Merel, S.; Zwiener, C.; Bräse, S.; Horn, H. Photolysis of four β-lactam antibiotics under simulated environmental conditions: Degradation, transformation products and antibacterial activity. *Sci. Total Environ.* **2019**, *651*, 1605–1612. [CrossRef]
50. Wang, X.H.; Lin, A.Y.C. Phototransformation of Cephalosporin Antibiotics in an Aqueous Environment Results in Higher Toxicity. *Environ. Sci. Technol.* **2012**, *46*, 12417–12426. [CrossRef]
51. Jiang, M.; Wang, L.; Ji, R. Biotic and abiotic degradation of four cephalosporin antibiotics in a lake surface water and sediment. *Chemosphere* **2010**, *80*, 1399–1405. [CrossRef]
52. Ribeiro, A.R.; Lutze, H.V.; Schmidt, T.C. Base-catalyzed hydrolysis and speciation-dependent photolysis of two cephalosporin antibiotics, ceftiofur and cefapirin. *Water Res.* **2018**, *134*, 253–260. [CrossRef]
53. Wei, X.; Chen, J.; Xie, Q.; Zhang, S.; Ge, L.; Qiao, X. Distinct photolytic mechanisms and products for different dissociation species of ciprofloxacin. *Environ. Sci. Technol.* **2013**, *47*, 4284–4290. [CrossRef]
54. Prabhakaran, D.; Sukul, P.; Lamshöft, M.; Maheswari, M.A.; Zühlke, S.; Spiteller, M. Photolysis of difloxacin and sarafloxacin in aqueous systems. *Chemosphere* **2009**, *77*, 739–746. [CrossRef]
55. Lastre-Acosta, A.M.; Barberato, B.; Parizi, M.P.S.; Teixeira, A.C.S.C. Direct and indirect photolysis of the antibiotic enoxacin: Kinetics of oxidation by reactive photo-induced species and simulations. *Environ. Sci. Pollut. Res.* **2019**, *26*, 4337–4347. [CrossRef] [PubMed]
56. Baena-Nogueras, R.M.; González-Mazo, E.; Lara-Martín, P.A. Photolysis of Antibiotics under Simulated Sunlight Irradiation: Identification of Photoproducts by High-Resolution Mass Spectrometry. *Environ. Sci. Technol.* **2017**, *51*, 3148–3156. [CrossRef] [PubMed]
57. Niu, J.; Zhang, L.; Li, Y.; Zhao, J.; Lv, S.; Xiao, K. Effects of environmental factors on sulfamethoxazole photodegradation under simulated sunlight irradiation: Kinetics and mechanism. *J. Environ. Sci.* **2013**, *25*, 1098–1106. [CrossRef]
58. López-Peñalver, J.J.; Sánchez-Polo, M.; Gómez-Pacheco, C.V.; Rivera-Utrilla, J. Photodegradation of tetracyclines in aqueous solution by using UV and UV/H_2O_2 oxidation processes. *J. Chem. Technol. Biotechnol.* **2010**, *85*, 1325–1333. [CrossRef]

59. Ryan, C.C.; Tan, D.T.; Arnold, W.A. Direct and indirect photolysis of sulfamethoxazole and trimethoprim in wastewater treatment plant effluent. *Water Res.* **2011**, *45*, 1280–1286. [CrossRef]
60. Shu, H.Y.; Chang, M.C. Decolorization and mineralization of a phthalocyanine dye C.I. Direct Blue 199 using UV/H_2O_2 process. *J. Hazard. Mater.* **2005**, *125*, 96–101. [CrossRef]
61. Wols, B.A.; Hofman-Caris, C.H.M. Review of photochemical reaction constants of organic micropollutants required for UV advanced oxidation processes in water. *Water Res.* **2012**, *46*, 2815–2827. [CrossRef]
62. Katsoyiannis, I.A.; Canonica, S.; Von Gunten, U. Efficiency and energy requirements for the transformation of organic micropollutants by ozone, O_3/H_2O_2 and UV/H_2O_2. *Water Res.* **2011**, *45*, 3811–3822. [CrossRef]
63. Andreozzi, R.; Canterino, M.; Marotta, R.; Paxeus, N. Antibiotic removal from wastewaters: The ozonation of amoxicillin. *J. Hazard. Mater.* **2005**, *122*, 243–250. [CrossRef]
64. Arslan-Alaton, I.; Dogruel, S. Pre-treatment of penicillin formulation effluent by advanced oxidation processes. *J. Hazard. Mater.* **2004**, *112*, 105–113. [CrossRef]
65. Alsager, O.A.; Alnajrani, M.N.; Abuelizz, H.A.; Aldaghmani, I.A. Removal of antibiotics from water and waste milk by ozonation: Kinetics, byproducts, and antimicrobial activity. *Ecotoxicol. Environ. Saf.* **2018**, *158*, 114–122. [CrossRef] [PubMed]
66. Witte, B.D.; Langenhove, H.V.; Hemelsoet, K.; Demeestere, K.; Wispelaere, P.D.; Van Speybroeck, V.; Dewulf, J. Levofloxacin ozonation in water: Rate determining process parameters and reaction pathway elucidation. *Chemosphere* **2009**, *76*, 683–689. [CrossRef] [PubMed]
67. Östman, M.; Björlenius, B.; Fick, J.; Tysklind, M. Effect of full-scale ozonation and pilot-scale granular activated carbon on the removal of biocides, antimycotics and antibiotics in a sewage treatment plant. *Sci. Total Environ.* **2019**, *649*, 1117–1123. [CrossRef] [PubMed]
68. Ternes, T.A.; Stüber, J.; Herrmann, N.; McDowell, D.; Ried, A.; Kampmann, M.; Teiser, B. Ozonation: A tool for removal of pharmaceuticals, contrast media and musk fragrances from wastewater? *Water Res.* **2003**, *37*, 1976–1982. [CrossRef]
69. Lange, F.; Cornelissen, S.; Kubac, D.; Sein, M.M.; Von Sonntag, J.; Hannich, C.B.; Golloch, A.; Heipieper, H.J.; Möder, M.; Von Sonntag, C. Degradation of macrolide antibiotics by ozone: A mechanistic case study with clarithromycin. *Chemosphere* **2006**, *65*, 17–23. [CrossRef]
70. Feng, M.; Yan, L.; Zhang, X.; Sun, P.; Yang, S.; Wang, L.; Wang, Z. Fast removal of the antibiotic flumequine from aqueous solution by ozonation: Influencing factors, reaction pathways, and toxicity evaluation. *Sci. Total Environ.* **2016**, *541*, 167–175. [CrossRef]
71. Qiang, Z.; Adams, C.; Surampalli, R. Determination of ozonation rate constants for lincomycin and spectinomycin. *Ozone Sci. Eng.* **2004**, *26*, 525–537. [CrossRef]
72. Andreozzi, R.; Canterino, M.; Giudice, R.L.; Marotta, R.; Pinto, G.; Pollio, A. Lincomycin solar photodegradation, algal toxicity and removal from wastewaters by means of ozonation. *Water Res.* **2006**, *40*, 630–638. [CrossRef]
73. Wang, H.; Mustafa, M.; Yu, G.; Östman, M.; Cheng, Y.; Wang, Y.; Tysklind, M. Oxidation of emerging biocides and antibiotics in wastewater by ozonation and the electro-peroxone process. *Chemosphere* **2019**, *235*, 575–585. [CrossRef]
74. Hirasawa, I.; Kaneko, S.; Kanai, Y.; Hosoya, S.; Okuyama, K.; Kamahara, T.; Adnan, A.; Mavinic, D.S.; Koch, F.A.; Villarroel Walker, R.; et al. Pilot-scale study of phosphorus recovery through struvite crystallization – examining the process feasibility. *J. Environ. Manag.* **2003**, *37*, 315–324.
75. Huber, M.M.; Canonica, S.; Park, G.Y.; Von Gunten, U. Oxidation of pharmaceuticals during ozonation and advanced oxidation processes. *Environ. Sci. Technol.* **2003**, *37*, 1016–1024. [CrossRef] [PubMed]
76. Garoma, T.; Umamaheshwar, S.K.; Mumper, A. Removal of sulfadiazine, sulfamethizole, sulfamethoxazole, and sulfathiazole from aqueous solution by ozonation. *Chemosphere* **2010**, *79*, 814–820. [CrossRef] [PubMed]
77. Dantas, R.F.; Contreras, S.; Sans, C.; Esplugas, S. Sulfamethoxazole abatement by means of ozonation. *J. Hazard. Mater.* **2008**, *150*, 790–794. [CrossRef] [PubMed]
78. Kim, T.-H.; Kim, S.D.; Kim, H.Y.; Lim, S.J.; Lee, M.; Yu, S. Degradation and toxicity assessment of sulfamethoxazole and chlortetracycline using electron beam, ozone and UV. *J. Hazard. Mater.* **2012**, *227*, 237–242. [CrossRef] [PubMed]
79. Chen, X.; Richard, J.; Liu, Y.; Dopp, E.; Tuerk, J.; Bester, K. Ozonation products of triclosan in advanced wastewater treatment. *Water Res.* **2012**, *46*, 2247–2256. [CrossRef]

80. Stylianou, S.K.; Katsoyiannis, I.A.; Mitrakas, M.; Zouboulis, A.I. Application of a ceramic membrane contacting process for ozone and peroxone treatment of micropollutant contaminated surface water. *J. Hazard. Mater.* **2018**, *358*, 129–135. [CrossRef]
81. Peyton, G.R.; Glaze, W.H. Destruction of pollutants in water with ozone in combination with ultraviolet radiation. 3. Photolysis of aqueous ozone. *Environ. Sci. Technol.* **1988**, *22*, 761–767. [CrossRef]
82. Litter, M.I.; Quici, N. Photochemical Advanced Oxidation Processes for Water and Wastewater Treatment. *Recent Patents Eng.* **2010**, *4*, 217–241. [CrossRef]
83. Morkovnik, A.F.; Okhlobystin, O.Y. Inorganic radical-ions and their organic reactions. *Russ. Chem. Rev.* **1979**, *48*, 1055–1075. [CrossRef]
84. Benitez, F.J.; Acero, J.L.; Real, F.J.; Roldan, G.; Casas, F. Comparison of different chemical oxidation treatments for the removal of selected pharmaceuticals in water matrices. *Chem. Eng. J.* **2011**, *168*, 1149–1156. [CrossRef]
85. Liu, P.; Zhang, H.; Feng, Y.; Yang, F.; Zhang, J. Removal of trace antibiotics from wastewater: A systematic study of nanofiltration combined with ozone-based advanced oxidation processes. *Chem. Eng. J.* **2014**, *240*, 211–220. [CrossRef]
86. Paucar, N.E.; Kim, I.; Tanaka, H.; Sato, C.; Paucar, N.E.; Kim, I.; Tanaka, H.; Sato, C. Effect of O_3 Dose on the O3/UV Treatment Process for the Removal of Pharmaceuticals and Personal Care Products in Secondary Effluent. *ChemEngineering* **2019**, *3*, 53. [CrossRef]
87. Lester, Y.; Avisar, D.; Gozlan, I.; Mamane, H. Removal of pharmaceuticals using combination of $UV/H_2O_2/O_3$ advanced oxidation process. *Water Sci. Technol.* **2011**, *64*, 2230–2238. [CrossRef] [PubMed]
88. Yao, W.; Ur Rehman, S.W.; Wang, H.; Yang, H.; Yu, G.; Wang, Y. Pilot-scale evaluation of micropollutant abatements by conventional ozonation, UV/O_3, and an electro-peroxone process. *Water Res.* **2018**, *138*, 106–117. [CrossRef] [PubMed]
89. Beltrán, F.J.; Aguinaco, A.; García-Araya, J.F.; Oropesa, A. Ozone and photocatalytic processes to remove the antibiotic sulfamethoxazole from water. *Water Res.* **2008**, *42*, 3799–3808. [CrossRef]
90. Glaze, W.H.; Beltran, F.; Tuhkanen, T.; Kang, J.W. Chemical models of advanced oxidation processes. *Water Pollut. Res. J. Can.* **1992**, *27*, 23–42. [CrossRef]
91. De Witte, B.; Dewulf, J.; Demeestere, K.; Van Langenhove, H. Ozonation and advanced oxidation by the peroxone process of ciprofloxacin in water. *J. Hazard. Mater.* **2009**, *161*, 701–708. [CrossRef]
92. Rosal, R.; Rodríguez, A.; Perdigón-Melón, J.A.; Mezcua, M.; Hernando, M.D.; Letón, P.; García-Calvo, E.; Agüera, A.; Fernández-Alba, A.R. Removal of pharmaceuticals and kinetics of mineralization by O_3/H_2O_2 in a biotreated municipal wastewater. *Water Res.* **2008**, *42*, 3719–3728. [CrossRef]
93. Gomes, D.S.; Gando-Ferreira, L.M.; Quinta-Ferreira, R.M.; Martins, R.C. Removal of sulfamethoxazole and diclofenac from water: Strategies involving O_3 and H_2O_2. *Environ. Technol.* **2018**, *39*, 1658–1669. [CrossRef]
94. Qin, W.; Song, Y.; Dai, Y.; Qiu, G.; Ren, M.; Zeng, P. Treatment of berberine hydrochloride pharmaceutical wastewater by $O_3/UV/H_2O_2$ advanced oxidation process. *Environ. Earth Sci.* **2015**, *73*, 4939–4946. [CrossRef]
95. Lee, H.; Lee, E.; Lee, C.H.; Lee, K. Degradation of chlorotetracycline and bacterial disinfection in livestock wastewater by ozone-based advanced oxidation. *J. Ind. Eng. Chem.* **2011**, *17*, 468–473. [CrossRef]
96. Luu, H.T.; Minh, D.N.; Lee, K. Effects of advanced oxidation of penicillin on biotoxicity, biodegradability and subsequent biological treatment. *Appl. Chem. Eng.* **2018**, *29*, 690–695.
97. Moradi, M.; Moussavi, G. Investigation of chemical-less UVC/VUV process for advanced oxidation of sulfamethoxazole in aqueous solutions: Evaluation of operational variables and degradation mechanism. *Sep. Purif. Technol.* **2018**, *190*, 90–99. [CrossRef]
98. Adams, C.; Wang, Y.; Loftin, K.; Meyer, M. Removal of antibiotics from surface and distilled water in conventional water treatment processes. *J. Environ. Eng.* **2002**, *128*, 253–260. [CrossRef]
99. Yuan, F.; Hu, C.; Hu, X.; Wei, D.; Chen, Y.; Qu, J. Photodegradation and toxicity changes of antibiotics in UV and UV/H_2O_2 process. *J. Hazard. Mater.* **2011**, *185*, 1256–1263. [CrossRef]
100. Jiao, S.; Zheng, S.; Yin, D.; Wang, L.; Chen, L. Aqueous photolysis of tetracycline and toxicity of photolytic products to luminescent bacteria. *Chemosphere* **2008**, *73*, 377–382. [CrossRef]
101. Baxendale, J.H.; Wilson, J.A. The photolysis of hydrogen peroxide at high light intensities. *Trans. Faraday Soc.* **1957**, *53*, 344–356. [CrossRef]
102. Goldstein, S.; Aschengrau, D.; Diamant, Y.; Rabani, J. Photolysis of Aqueous H_2O_2: Quantum Yield and Applications for Polychromatic UV Actinometry in Photoreactors. *Environ. Sci. Technol.* **2007**, *41*, 7486–7490. [CrossRef]

103. Cédat, B.; De Brauer, C.; Métivier, H.; Dumont, N.; Tutundjan, R. Are UV photolysis and UV/H_2O_2 process efficient to treat estrogens in waters? Chemical and biological assessment at pilot scale. *Water Res.* **2016**, *100*, 357–366. [CrossRef]
104. Ross, F.; Ross, A.B. Selected specific rates of reactions of transients from water in aqueous solution. iii. hydroxyl radical and perhydroxyl radical and their radical ions. *Natl. Bur. Stand. Natl. Stand. Ref. Data Ser.* **1977**. [CrossRef]
105. Bielski, B.H.J.; Cabelli, D.E.; Arudi, R.L.; Ross, A.B. Reactivity of HO_2/O_2 Radicals in Aqueous Solution. *J. Phys. Chem. Ref. Data* **1985**, *14*, 1041–1100. [CrossRef]
106. Christensen, H.; Sehested, K.; Corfitzen, H. Reactions of hydroxyl radicals with hydrogen peroxide at ambient and elevated temperatures. *J. Phys. Chem.* **1982**, *86*, 1588–1590. [CrossRef]
107. Jung, Y.J.; Kim, W.G.; Yoon, Y.; Kang, J.W.; Hong, Y.M.; Kim, H.W. Removal of amoxicillin by UV and UV/H_2O_2 processes. *Sci. Total Environ.* **2012**, *420*, 160–167. [CrossRef] [PubMed]
108. Sun, Y.; Cho, D.W.; Graham, N.J.D.; Hou, D.; Yip, A.C.K.; Khan, E.; Song, H.; Li, Y.; Tsang, D.C.W. Degradation of antibiotics by modified vacuum-UV based processes: Mechanistic consequences of H_2O_2 and $K_2S_2O_8$ in the presence of halide ions. *Sci. Total Environ.* **2019**, *664*, 312–321. [CrossRef]
109. Khorsandi, H.; Teymori, M.; Aghapour, A.A.; Jafari, S.J.; Taghipour, S.; Bargeshadi, R. Photodegradation of ceftriaxone in aqueous solution by using UVC and UVC/H_2O_2 oxidation processes. *Appl. Water Sci.* **2019**, *9*, 81. [CrossRef]
110. De Souza Santos, L.V.; Meireles, A.M.; Lange, L.C. Degradation of antibiotics norfloxacin by Fenton, UV and UV/H_2O_2. *J. Environ. Manag.* **2015**, *154*, 8–12. [CrossRef]
111. Urbano, V.R.; Peres, M.S.; Maniero, M.G.; Guimarães, J.R. Abatement and toxicity reduction of antimicrobials by UV/H_2O_2 process. *J. Environ. Manag.* **2017**, *193*, 439–447. [CrossRef]
112. Li, W.; Xu, X.; Lyu, B.; Tang, Y.; Zhang, Y.; Chen, F.; Korshin, G. Degradation of typical macrolide antibiotic roxithromycin by hydroxyl radical: Kinetics, products, and toxicity assessment. *Environ. Sci. Pollut. Res.* **2019**. [CrossRef]
113. Zhang, R.; Yang, Y.; Huang, C.H.; Zhao, L.; Sun, P. Kinetics and modeling of sulfonamide antibiotic degradation in wastewater and human urine by UV/H_2O_2 and UV/PDS. *Water Res.* **2016**, *103*, 283–292. [CrossRef]
114. Lin, C.C.; Wu, M.S. Feasibility of using UV/H_2O_2 process to degrade sulfamethazine in aqueous solutions in a large photoreactor. *J. Photochem. Photobiol. A Chem.* **2018**, *367*, 446–451. [CrossRef]
115. Lekkerkerker-Teunissen, K.; Benotti, M.J.; Snyder, S.A.; Van Dijk, H.C. Transformation of atrazine, carbamazepine, diclofenac and sulfamethoxazole by low and medium pressure UV and UV/H_2O_2 treatment. *Sep. Purif. Technol.* **2012**, *96*, 33–43. [CrossRef]
116. Navalon, S.; Alvaro, M.; Garcia, H. Heterogeneous Fenton catalysts based on clays, silicas and zeolites. *Appl. Catal. B Environ.* **2010**, *99*, 1–26. [CrossRef]
117. Minero, C.; Lucchiari, M.; Vione, D.; Maurino, V. Fe(III)-enhanced sonochemical degradation of methylene blue in aqueous solution. *Environ. Sci. Technol.* **2005**, *39*, 8936–8942. [CrossRef] [PubMed]
118. Walling, C. Fenton's Reagent Revisited. *Acc. Chem. Res.* **1975**, *8*, 125–131. [CrossRef]
119. Buxton, G.V.; Greenstock, C.L.; Helman, W.P.; Ross, A.B. Critical Review of rate constants for reactions of hydrated electrons, hydrogen atoms and hydroxyl radicals (·OH/·O−in Aqueous Solution. *J. Phys. Chem. Ref. Data* **1988**, *17*, 513–886. [CrossRef]
120. Kulik, N.; Trapido, M.; Goi, A.; Veressinina, Y.; Munter, R. Combined chemical treatment of pharmaceutical effluents from medical ointment production. *Chemosphere* **2008**, *70*, 1525–1531. [CrossRef]
121. Ay, F.; Kargi, F. Advanced oxidation of amoxicillin by Fenton's reagent treatment. *J. Hazard. Mater.* **2010**, *179*, 622–627. [CrossRef]
122. Homem, V.; Alves, A.; Santos, L. Amoxicillin degradation at ppb levels by Fenton's oxidation using design of experiments. *Sci. Total Environ.* **2010**, *408*, 6272–6280. [CrossRef]
123. Türkay, G.K.; Kumbur, H. Investigation of amoxicillin removal from aqueous solution by Fenton and photocatalytic oxidation processes. *Kuwait J. Sci.* **2019**, *46*, 85–93.
124. Rozas, O.; Contreras, D.; Mondaca, M.A.; Pérez-Moya, M.; Mansilla, H.D. Experimental design of Fenton and photo-Fenton reactions for the treatment of ampicillin solutions. *J. Hazard. Mater.* **2010**, *177*, 1025–1030. [CrossRef]

125. Pulicharla, R.; Brar, S.K.; Rouissi, T.; Auger, S.; Drogui, P.; Verma, M.; Surampalli, R.Y. Degradation of chlortetracycline in wastewater sludge by ultrasonication, Fenton oxidation, and ferro-sonication. *Ultrason. Sonochem.* **2017**, *34*, 332–342. [CrossRef] [PubMed]
126. Salari, M.; Rakhshandehroo, G.R.; Nikoo, M.R. Degradation of ciprofloxacin antibiotic by Homogeneous Fenton oxidation: Hybrid AHP-PROMETHEE method, optimization, biodegradability improvement and identification of oxidized by-products. *Chemosphere* **2018**, *206*, 157–167. [CrossRef] [PubMed]
127. Rakhshandehroo, G.R.; Salari, M.; Nikoo, M.R. Optimization of degradation of ciprofloxacin antibiotic and assessment of degradation products using full factorial experimental design by fenton homogenous process. *Glob. Nest J.* **2018**, *20*, 324–332.
128. Gupta, A.; Garg, A. Degradation of ciprofloxacin using Fenton's oxidation: Effect of operating parameters, identification of oxidized by-products and toxicity assessment. *Chemosphere* **2018**, *193*, 1181–1188. [CrossRef]
129. Borghi, A.A.; Silva, M.F.; Al Arni, S.; Converti, A.; Palma, M.S.A. Doxycycline degradation by the oxidative Fenton process. *J. Chem.* **2015**, *2015*, 492030. [CrossRef]
130. Rodrigues-Silva, C.; Maniero, M.G.; Rath, S.; Guimarães, J.R. Degradation of flumequine by the Fenton and photo-Fenton processes: Evaluation of residual antimicrobial activity. *Sci. Total Environ.* **2013**, *445–446*, 337–346. [CrossRef]
131. Epold, I.; Trapido, M.; Dulova, N. Degradation of levofloxacin in aqueous solutions by Fenton, ferrous ion-activated persulfate and combined Fenton/persulfate systems. *Chem. Eng. J.* **2015**, *279*, 452–462. [CrossRef]
132. Wang, S.; Wang, J. Comparative study on sulfamethoxazole degradation by Fenton and Fe(ii)-activated persulfate process. *RSC Adv.* **2017**, *7*, 48670–48677. [CrossRef]
133. Wang, S.; Wang, J. Trimethoprim degradation by Fenton and Fe(II)-activated persulfate processes. *Chemosphere* **2018**, *191*, 97–105. [CrossRef]
134. Gallard, H.; De Laat, J. Kinetic modelling of Fe(III)/H_2O_2 oxidation reactions in dilute aqueous solution using atrazine as a model organic compound. *Water Res.* **2000**, *34*, 3107–3116. [CrossRef]
135. Neyens, E.; Baeyens, J. A review of classic Fenton's peroxidation as an advanced oxidation technique. *J. Hazard. Mater.* **2003**, *98*, 33–50. [CrossRef]
136. Zhang, N.; Chen, J.; Fang, Z.; Tsang, E.P. Ceria accelerated nanoscale zerovalent iron assisted heterogenous Fenton oxidation of tetracycline. *Chem. Eng. J.* **2019**, *369*, 588–599. [CrossRef]
137. Zhuan, R.; Wang, J. Enhanced mineralization of sulfamethoxazole by gamma radiation in the presence of Fe_3O_4 as Fenton-like catalyst. *Environ. Sci. Pollut. Res.* **2019**, *26*, 27712–27725. [CrossRef] [PubMed]
138. Wang, A.; Wang, H.; Deng, H.; Wang, S.; Shi, W.; Yi, Z.; Qiu, R.; Yan, K. Controllable synthesis of mesoporous manganese oxide microsphere efficient for photo-Fenton-like removal of fluoroquinolone antibiotics. *Appl. Catal. B Environ.* **2019**, *248*, 298–308. [CrossRef]
139. Hu, Y.; Chen, K.; Li, Y.L.; He, J.Y.; Zhang, K.S.; Liu, T.; Xu, W.; Huang, X.J.; Kong, L.T.; Liu, J.H. Morphology-tunable WMoO nanowire catalysts for the extremely efficient elimination of tetracycline: Kinetics, mechanisms and intermediates. *Nanoscale* **2019**, *11*, 1047–1057. [CrossRef] [PubMed]
140. Wang, Q.; Ma, Y.; Xing, S. Comparative study of Cu-based bimetallic oxides for Fenton-like degradation of organic pollutants. *Chemosphere* **2018**, *203*, 450–456. [CrossRef] [PubMed]
141. Wang, R.Z.; Huang, D.L.; Liu, Y.G.; Zhang, C.; Lai, C.; Wang, X.; Zeng, G.M.; Gong, X.M.; Duan, A.; Zhang, Q.; et al. Recent advances in biochar-based catalysts: Properties, applications and mechanisms for pollution remediation. *Chem. Eng. J.* **2019**, *371*, 380–403. [CrossRef]
142. Xu, L.; Yang, Y.; Li, W.; Tao, Y.; Sui, Z.; Song, S.; Yang, J. Three-dimensional macroporous graphene-wrapped zero-valent copper nanoparticles as efficient micro-electrolysis-promoted Fenton-like catalysts for metronidazole removal. *Sci. Total Environ.* **2019**, *658*, 219–233. [CrossRef]
143. Zha, S.; Cheng, Y.; Gao, Y.; Chen, Z.; Megharaj, M.; Naidu, R. Nanoscale zero-valent iron as a catalyst for heterogeneous Fenton oxidation of amoxicillin. *Chem. Eng. J.* **2014**, *255*, 141–148. [CrossRef]
144. Mondal, S.K.; Saha, A.K.; Sinha, A. Removal of ciprofloxacin using modified advanced oxidation processes: Kinetics, pathways and process optimization. *J. Clean. Prod.* **2018**, *171*, 1203–1214. [CrossRef]
145. Li, J.; Pan, L.; Yu, G.; Xie, S.; Li, C.; Lai, D.; Li, Z.; You, F.; Wang, Y. The synthesis of heterogeneous Fenton-like catalyst using sewage sludge biochar and its application for ciprofloxacin degradation. *Sci. Total Environ.* **2019**, *654*, 1284–1292. [CrossRef] [PubMed]

146. Fang, Z.; Chen, J.; Qiu, X.; Qiu, X.; Cheng, W.; Zhu, L. Effective removal of antibiotic metronidazole from water by nanoscale zero-valent iron particles. *Desalination* **2011**, *268*, 60–67. [CrossRef]
147. Huang, D.; Luo, H.; Zhang, C.; Zeng, G.; Lai, C.; Cheng, M.; Wang, R.; Deng, R.; Xue, W.; Gong, X.; et al. Nonnegligible role of biomass types and its compositions on the formation of persistent free radicals in biochar: Insight into the influences on Fenton-like process. *Chem. Eng. J.* **2019**, *361*, 353–363. [CrossRef]
148. Zhuang, Y.; Liu, Q.; Kong, Y.; Shen, C.; Hao, H.; Dionysiou, D.D.; Shi, B. Enhanced antibiotic removal through a dual-reaction-center Fenton-like process in 3D graphene based hydrogels. *Environ. Sci. Nano* **2019**, *6*, 388–398. [CrossRef]
149. Sedlak, D.L.; Andren, A.W. Oxidation of chlorobenzene with Fenton's reagent. *Environ. Sci. Technol.* **1991**, *25*, 777–782. [CrossRef]
150. Pignatello, J.J. Dark and photoassisted iron(3+)-catalyzed degradation of chlorophenoxy herbicides by hydrogen peroxide. *Environ. Sci. Technol.* **1992**, *26*, 944–951. [CrossRef]
151. Bossmann, S.H.; Oliveros, E.; Göb, S.; Siegwart, S.; Dahlen, E.P.; Payawan, L.; Straub, M.; Wörner, M.; Braun, A.M. New Evidence against Hydroxyl Radicals as Reactive Intermediates in the Thermal and Photochemically Enhanced Fenton Reactions. *J. Phys. Chem. A* **1998**, *102*, 5542–5550. [CrossRef]
152. Hinojosa Guerra, M.M.; Oller Alberola, I.; Malato Rodriguez, S.; Agüera López, A.; Acevedo Merino, A.; Quiroga Alonso, J.M. Oxidation mechanisms of amoxicillin and paracetamol in the photo-Fenton solar process. *Water Res.* **2019**, *156*, 232–240. [CrossRef]
153. Ioannou-Ttofa, L.; Raj, S.; Prakash, H.; Fatta-Kassinos, D. Solar photo-Fenton oxidation for the removal of ampicillin, total cultivable and resistant E. coli and ecotoxicity from secondary-treated wastewater effluents. *Chem. Eng. J.* **2019**, *355*, 91–102. [CrossRef]
154. Amildon Ricardo, I.; Paiva, V.A.B.; Paniagua, C.E.S.; Trovó, A.G. Chloramphenicol photo-Fenton degradation and toxicity changes in both surface water and a tertiary effluent from a municipal wastewater treatment plant at near-neutral conditions. *Chem. Eng. J.* **2018**, *347*, 763–770. [CrossRef]
155. Giri, A.S.; Golder, A.K. Ciprofloxacin degradation in photo-Fenton and photo-catalytic processes: Degradation mechanisms and iron chelation. *J. Environ. Sci.* **2019**, *80*, 82–92. [CrossRef] [PubMed]
156. Giraldo-Aguirre, A.L.; Serna-Galvis, E.A.; Erazo-Erazo, E.D.; Silva-Agredo, J.; Giraldo-Ospina, H.; Flórez-Acosta, O.A.; Torres-Palma, R.A. Removal of β-lactam antibiotics from pharmaceutical wastewaters using photo-Fenton process at near-neutral pH. *Environ. Sci. Pollut. Res.* **2018**, *25*, 20293–20303. [CrossRef] [PubMed]
157. Wang, Q.; Pang, W.; Mao, Y.; Sun, Q.; Zhang, P.; Ke, Q.; Yu, H.; Dai, C.; Zhao, M. Study of the degradation of trimethoprim using photo-Fenton oxidation technology. *Water* **2019**, *11*, 207. [CrossRef]
158. Gogate, P.R.; Pandit, A.B. A review of imperative technologies for wastewater treatment I: Oxidation technologies at ambient conditions. *Adv. Environ. Res.* **2004**, *8*, 501–551. [CrossRef]
159. Gogate, P.R.; Pandit, A.B. A review of imperative technologies for wastewater treatment II: Hybrid methods. *Adv. Environ. Res.* **2004**, *8*, 553–597. [CrossRef]
160. Pirkanniemi, K.; Sillanpää, M. Heterogeneous water phase catalysis as an environmental application: A review. *Chemosphere* **2002**, *48*, 1047–1060. [CrossRef]
161. Wang, N.; Zheng, T.; Zhang, G.; Wang, P. A review on Fenton-like processes for organic wastewater treatment. *J. Environ. Chem. Eng.* **2016**, *4*, 762–787. [CrossRef]
162. Hao, H.; Shi, J.L.; Xu, H.; Li, X.; Lang, X. N-hydroxyphthalimide-TiO_2 complex visible light photocatalysis. *Appl. Catal. B Environ.* **2019**, *246*, 149–155. [CrossRef]
163. Méndez-Medrano, M.G.; Kowalska, E.; Lehoux, A.; Herissan, A.; Ohtani, B.; Bahena, D.; Briois, V.; Colbeau-Justin, C.; Rodríguez-López, J.L.; Remita, H. Surface Modification of TiO_2 with Ag Nanoparticles and CuO Nanoclusters for Application in Photocatalysis. *J. Phys. Chem. C* **2016**, *120*, 5143–5154. [CrossRef]
164. Sanzone, G.; Zimbone, M.; Cacciato, G.; Ruffino, F.; Carles, R.; Privitera, V.; Grimaldi, M.G. Ag/TiO_2 nanocomposite for visible light-driven photocatalysis. *Superlattices Microstruct.* **2018**, *123*, 394–402. [CrossRef]
165. Xiong, X.; Xu, Y. Synergetic Effect of Pt and Borate on the TiO_2-Photocatalyzed Degradation of Phenol in Water. *J. Phys. Chem. C* **2016**, *120*, 3906–3912. [CrossRef]
166. Wang, D.; Wang, S.; Li, B.; Zhang, Z.; Zhang, Q. Tunable band gap of NV co-doped Ca:TiO_2B ($CaTi_5O_{11}$) for visible-light photocatalysis. *Int. J. Hydrogen Energy* **2019**, *44*, 4716–4723. [CrossRef]

167. Du, D.; Shi, W.; Wang, L.; Zhang, J. Yolk-shell structured Fe_3O_4@void@TiO_2 as a photo-Fenton-like catalyst for the extremely efficient elimination of tetracycline. *Appl. Catal. B Environ.* **2017**, *200*, 484–492. [CrossRef]
168. Gurkan, Y.Y.; Kasapbasi, E.; Cinar, Z. Enhanced solar photocatalytic activity of TiO_2 by selenium(IV) ion-doping: Characterization and DFT modeling of the surface. *Chem. Eng. J.* **2013**, *214*, 34–44. [CrossRef]
169. Eswar, N.K.; Ramamurthy, P.C.; Madras, G. Novel synergistic photocatalytic degradation of antibiotics and bacteria using V–N doped TiO_2 under visible light: The state of nitrogen in V-doped TiO_2. *New J. Chem.* **2016**, *40*, 3464–3475. [CrossRef]
170. Zheng, X.; Xu, S.; Wang, Y.; Sun, X.; Gao, Y.; Gao, B. Enhanced degradation of ciprofloxacin by graphitized mesoporous carbon (GMC)-TiO_2 nanocomposite: Strong synergy of adsorption-photocatalysis and antibiotics degradation mechanism. *J. Colloid Interface Sci.* **2018**, *527*, 202–213. [CrossRef]
171. Serna-Galvis, E.A.; Giraldo-Aguirre, A.L.; Silva-Agredo, J.; Flórez-Acosta, O.A.; Torres-Palma, R.A. Removal of antibiotic cloxacillin by means of electrochemical oxidation, TiO_2 photocatalysis, and photo-Fenton processes: Analysis of degradation pathways and effect of the water matrix on the elimination of antimicrobial activity. *Environ. Sci. Pollut. Res.* **2017**, *24*, 6339–6352. [CrossRef]
172. Tran, M.L.; Fu, C.C.; Juang, R.S. Effects of water matrix components on degradation efficiency and pathways of antibiotic metronidazole by UV/TiO_2 photocatalysis. *J. Mol. Liq.* **2019**, *276*, 32–38. [CrossRef]
173. Shankaraiah, G.; Poodari, S.; Bhagawan, D.; Himabindu, V.; Vidyavathi, S. Degradation of antibiotic norfloxacin in aqueous solution using advanced oxidation processes (AOPs)—A comparative study. *Desalin. Water Treat.* **2016**, *57*, 27804–27815. [CrossRef]
174. Serna-Galvis, E.A.; Silva-Agredo, J.; Giraldo, A.L.; Flórez-Acosta, O.A.; Torres-Palma, R.A. Comparative study of the effect of pharmaceutical additives on the elimination of antibiotic activity during the treatment of oxacillin in water by the photo-Fenton, TiO_2-photocatalysis and electrochemical processes. *Sci. Total Environ.* **2016**, *541*, 1431–1438. [CrossRef]
175. Jiang, Q.; Zhu, R.; Zhu, Y.; Chen, Q. Efficient degradation of cefotaxime by a UV+ferrihydrite/TiO_2+H_2O_2 process: The important role of ferrihydrite in transferring photo-generated electrons from TiO_2 to H_2O_2. *J. Chem. Technol. Biotechnol.* **2019**, *94*, 2512–2521. [CrossRef]
176. García-Muñoz, P.; Zussblatt, N.P.; Pliego, G.; Zazo, J.A.; Fresno, F.; Chmelka, B.F.; Casas, J.A. Evaluation of photoassisted treatments for norfloxacin removal in water using mesoporous Fe_2O_3-TiO_2 materials. *J. Environ. Manag.* **2019**, *238*, 243–250. [CrossRef] [PubMed]
177. Pokhrel, N.; Vabbina, P.K.; Pala, N. Sonochemistry: Science and Engineering. *Ultrason. Sonochem.* **2016**, *29*, 104–128. [CrossRef] [PubMed]
178. González-García, J.; Sáez, V.; Tudela, I.; Díez-Garcia, M.I.; Deseada Esclapez, M.; Louisnard, O. Sonochemical Treatment of Water Polluted by Chlorinated Organocompounds. A Review. *Water* **2010**, *2*, 28–74. [CrossRef]
179. Hao, H.; Chen, Y.; Wu, M.; Wang, H.; Yin, Y.; Lü, Z. Sonochemistry of degrading p-chlorophenol in water by high frequency ultrasound. *Ultrason. Sonochem.* **2004**, *11*, 43–46. [CrossRef]
180. Sathishkumar, P.; Mangalaraja, R.V.; Anandan, S. Review on the recent improvements in sonochemical and combined sonochemical oxidation processes—A powerful tool for destruction of environmental contaminants. *Renew. Sustain. Energy Rev.* **2016**, *55*, 426–454. [CrossRef]
181. Serna-Galvis, E.A.; Botero-Coy, A.M.; Martínez-Pachón, D.; Moncayo-Lasso, A.; Ibáñez, M.; Hernández, F.; Torres-Palma, R.A. Degradation of seventeen contaminants of emerging concern in municipal wastewater effluents by sonochemical advanced oxidation processes. *Water Res.* **2019**, *154*, 349–360. [CrossRef]
182. Bagal, M.V.; Gogate, P.R. Wastewater treatment using hybrid treatment schemes based on cavitation and Fenton chemistry: A review. *Ultrason. Sonochem.* **2014**, *21*, 1–14. [CrossRef]
183. Dindarsafa, M.; Khataee, A.; Kaymak, B.; Vahid, B.; Karimi, A.; Rahmani, A. Heterogeneous sono-Fenton-like process using martite nanocatalyst prepared by high energy planetary ball milling for treatment of a textile dye. *Ultrason. Sonochem.* **2017**, *34*, 389–399. [CrossRef]
184. Zhou, T.; Wu, X.; Zhang, Y.; Li, J.; Lim, T.T. Synergistic catalytic degradation of antibiotic sulfamethazine in a heterogeneous sonophotolytic goethite/oxalate Fenton-like system. *Appl. Catal. B Environ.* **2013**, *136-137*, 294–301. [CrossRef]
185. Ghoreishian, S.M.; Raju, G.S.R.; Pavitra, E.; Kwak, C.H.; Han, Y.K.; Huh, Y.S. Ultrasound-assisted heterogeneous degradation of tetracycline over flower-like rGO/CdWO4 hierarchical structures as robust

solar-light-responsive photocatalysts: Optimization, kinetics, and mechanism. *Appl. Surf. Sci.* **2019**, *489*, 110–122. [CrossRef]
186. Vinesh, V.; Shaheer, A.R.M.; Neppolian, B. Reduced graphene oxide (rGO) supported electron deficient B-doped TiO_2 (Au/B-TiO_2/rGO) nanocomposite: An efficient visible light sonophotocatalyst for the degradation of Tetracycline (TC). *Ultrason. Sonochem.* **2019**, *50*, 302–310. [CrossRef] [PubMed]
187. Abazari, R.; Mahjoub, A.R.; Sanati, S.; Rezvani, Z.; Hou, Z.; Dai, H. Ni-Ti Layered Double Hydroxide@Graphitic Carbon Nitride Nanosheet: A Novel Nanocomposite with High and Ultrafast Sonophotocatalytic Performance for Degradation of Antibiotics. *Inorg. Chem.* **2019**, *58*, 1834–1849. [CrossRef] [PubMed]
188. Brillas, E.; Baños, M.Á.; Skoumal, M.; Cabot, P.L.; Garrido, J.A.; Rodríguez, R.M. Degradation of the herbicide 2,4-DP by anodic oxidation, electro-Fenton and photoelectro-Fenton using platinum and boron-doped diamond anodes. *Chemosphere* **2007**, *68*, 199–209. [CrossRef] [PubMed]
189. Mirzaei, S.; Javanbakht, V. Dye removal from aqueous solution by a novel dual cross-linked biocomposite obtained from mucilage of Plantago Psyllium and eggshell membrane. *Int. J. Biol. Macromol.* **2019**, *134*, 1187–1204. [CrossRef]
190. Panizza, M.; Delucchi, M.; Cerisola, G. Electrochemical degradation of anionic surfactants. *J. Appl. Electrochem.* **2005**, *35*, 357–361. [CrossRef]
191. Panizza, M.; Cerisola, G. Direct And Mediated Anodic Oxidation of Organic Pollutants. *Chem. Rev.* **2009**, *109*, 6541–6569. [CrossRef]
192. Cañizares, P.; Sáez, C.; Sánchez-Carretero, A.; Rodrigo, M.A. Synthesis of novel oxidants by electrochemical technology. *J. Appl. Electrochem.* **2009**, *39*, 2143. [CrossRef]
193. Rodrigo, M.A.; Cañizares, P.; Sánchez-Carretero, A.; Sáez, C. Use of conductive-diamond electrochemical oxidation for wastewater treatment. *Catal. Today* **2010**, *151*, 173–177. [CrossRef]
194. Cañizares, P.; Sáez, C.; Sánchez-Carretero, A.; Rodrigo, M.A. Influence of the characteristics of p-Si BDD anodes on the efficiency of peroxodiphosphate electrosynthesis process. *Electrochem. Commun.* **2008**, *10*, 602–606. [CrossRef]
195. Comninellis, C. Electrocatalysis in the electrochemical conversion/combustion of organic pollutants for waste water treatment. *Electrochim. Acta* **1994**, *39*, 1857–1862. [CrossRef]
196. Waterston, K.; Wang, J.W.; Bejan, D.; Bunce, N.J. Electrochemical waste water treatment: Electrooxidation of acetaminophen. *J. Appl. Electrochem.* **2006**, *36*, 227–232. [CrossRef]
197. Yosypchuk, O.; Barek, J.; Vyskočil, V. Voltammetric Determination of Carcinogenic Derivatives of Pyrene Using a Boron-Doped Diamond Film Electrode. *Anal. Lett.* **2012**, *45*, 449–459. [CrossRef]
198. Lima, A.B.; Faria, E.O.; Montes, R.H.O.; Cunha, R.R.; Richter, E.M.; Munoz, R.A.A.; dos Santos, W.T.P. Electrochemical Oxidation of Ibuprofen and Its Voltammetric Determination at a Boron-Doped Diamond Electrode. *Electroanalysis* **2013**, *25*, 1585–1588. [CrossRef]
199. Pereira, P.F.; Marra, M.C.; Lima, A.B.; Dos Santos, W.T.P.; Munoz, R.A.A.; Richter, E.M. Fast and simultaneous determination of nimesulide and paracetamol by batch injection analysis with amperometric detection on bare boron-doped diamond electrode. *Diam. Relat. Mater.* **2013**, *39*, 41–46. [CrossRef]
200. Sun, H.; Dong, L.; Yu, H.; Huo, M. Direct electrochemical oxidation and detection of hydrazine on a boron doped diamond (BDD) electrode. *Russ. J. Electrochem.* **2013**, *49*, 883–887. [CrossRef]
201. Zou, Y.S.; He, L.L.; Zhang, Y.C.; Shi, X.Q.; Li, Z.X.; Zhou, Y.L.; Tu, C.J.; Gu, L.; Zeng, H.B. The microstructure and electrochemical properties of boron-doped nanocrystalline diamond film electrodes and their application in non-enzymatic glucose detection. *J. Appl. Electrochem.* **2013**, *43*, 911–917. [CrossRef]
202. Marselli, B.; Garcia-Gomez, J.; Michaud, P.A.; Rodrigo, M.A.; Comninellis, C. Electrogeneration of hydroxyl radicals on boron-doped diamond electrodes. *J. Electrochem. Soc.* **2003**, *150*, D79–D83. [CrossRef]
203. Sun, J.; Lu, H.; Du, L.; Lin, H.; Li, H. Anodic oxidation of anthraquinone dye Alizarin Red S at Ti/BDD electrodes. *Appl. Surf. Sci.* **2011**, *257*, 6667–6671. [CrossRef]
204. Migliorini, F.L.; Braga, N.A.; Alves, S.A.; Lanza, M.R.V.; Baldan, M.R.; Ferreira, N.G. Anodic oxidation of wastewater containing the Reactive Orange 16 Dye using heavily boron-doped diamond electrodes. *J. Hazard. Mater.* **2011**, *192*, 1683–1689. [CrossRef]
205. Iniesta, J.; Michaud, P.A.; Panizza, M.; Cerisola, G.; Aldaz, A.; Comninellis, C. Electrochemical oxidation of phenol at boron-doped diamond electrode. *Electrochim. Acta* **2001**, *46*, 3573–3578. [CrossRef]

206. Hupert, M.; Muck, A.; Wang, J.; Stotter, J.; Cvackova, Z.; Haymond, S.; Show, Y.; Swain, G.M. Conductive diamond thin-films in electrochemistry. *Diam. Relat. Mater.* **2003**, *12*, 1940–1949. [CrossRef]
207. Boye, B.; Brillas, E.; Marselli, B.; Michaud, P.A.; Comninellis, C.; Farnia, G.; Sandonà, G. Electrochemical incineration of chloromethylphenoxy herbicides in acid medium by anodic oxidation with boron-doped diamond electrode. *Electrochim. Acta* **2006**, *51*, 2872–2880. [CrossRef]
208. Cañizares, P.; Gadri, A.; Lobato, J.; Nasr, B.; Paz, R.; Rodrigo, M.A.; Saez, C. Electrochemical Oxidation of Azoic Dyes with Conductive-Diamond Anodes. *Ind. Eng. Chem. Res.* **2006**, *45*, 3468–3473. [CrossRef]
209. Oturan, N.; Hamza, M.; Ammar, S.; Abdelhédi, R.; Oturan, M.A. Oxidation/mineralization of 2-Nitrophenol in aqueous medium by electrochemical advanced oxidation processes using Pt/carbon-felt and BDD/carbon-felt cells. *J. Electroanal. Chem.* **2011**, *661*, 66–71. [CrossRef]
210. El-Ghenymy, A.; Arias, C.; Cabot, P.L.; Centellas, F.; Garrido, J.A.; Rodríguez, R.M.; Brillas, E. Electrochemical incineration of sulfanilic acid at a boron-doped diamond anode. *Chemosphere* **2012**, *87*, 1126–1133. [CrossRef]
211. Cavalcanti, E.B.; Segura, S.G.; Centellas, F.; Brillas, E. Electrochemical incineration of omeprazole in neutral aqueous medium using a platinum or boron-doped diamond anode: Degradation kinetics and oxidation products. *Water Res.* **2013**, *47*, 1803–1815. [CrossRef]
212. Rabaaoui, N.; Moussaoui, Y.; Allagui, M.S.; Ahmed, B.; Elaloui, E. Anodic oxidation of nitrobenzene on BDD electrode: Variable effects and mechanisms of degradation. *Sep. Purif. Technol.* **2013**, *107*, 318–323. [CrossRef]
213. Sales Solano, A.M.; Costa De Araújo, C.K.; Vieira De Melo, J.; Peralta-Hernandez, J.M.; Ribeiro Da Silva, D.; Martínez-Huitle, C.A. Decontamination of real textile industrial effluent by strong oxidant species electrogenerated on diamond electrode: Viability and disadvantages of this electrochemical technology. *Appl. Catal. B Environ.* **2013**, *130*, 112–120. [CrossRef]
214. Flox, C.; Cabot, P.L.; Centellas, F.; Garrido, J.A.; Rodríguez, R.M.; Arias, C.; Brillas, E. Electrochemical combustion of herbicide mecoprop in aqueous medium using a flow reactor with a boron-doped diamond anode. *Chemosphere* **2006**, *64*, 892–902. [CrossRef]
215. Bergmann, M.E.H.; Rollin, J.; Iourtchouk, T. The occurrence of perchlorate during drinking water electrolysis using BDD anodes. *Electrochim. Acta* **2009**, *54*, 2102–2107. [CrossRef]
216. Sánchez-Carretero, A.; Sáez, C.; Cañizares, P.; Rodrigo, M.A. Electrochemical production of perchlorates using conductive diamond electrolyses. *Chem. Eng. J.* **2011**, *166*, 710–714. [CrossRef]
217. Rodríguez-Chueca, J.; Varella della Giustina, S.; Rocha, J.; Fernandes, T.; Pablos, C.; Encinas, Á.; Barceló, D.; Rodríguez-Mozaz, S.; Manaia, C.M.; Marugán, J. Assessment of full-scale tertiary wastewater treatment by UV-C based-AOPs: Removal or persistence of antibiotics and antibiotic resistance genes? *Sci. Total Environ.* **2019**, *652*, 1051–1061. [CrossRef] [PubMed]
218. Chowdhury, P.; Sarathy, S.R.; Das, S.; Li, J.; Ray, A.K.; Ray, M.B. Direct UV photolysis of pharmaceutical compounds: Determination of pH-dependent quantum yield and full-scale performance. *Chem. Eng. J.* **2020**, *380*, 122460. [CrossRef]
219. Rame; Tridecima, A.; Pranoto, H. FLASH Technology: Full-Scale Hospital Waste Water Treatments Adopted in Aceh. In Proceedings of the E3S Web of Conferences, Semarang, Indonesia, 15–16 August 2017; Volume 31.
220. Sui, Q.; Huang, J.; Lu, S.; Deng, S.; Wang, B.; Zhao, W.; Qiu, Z.; Yu, G. Removal of pharmaceutical and personal care products by sequential ultraviolet and ozonation process in a full-scale wastewater treatment plant. *Front. Environ. Sci. Eng.* **2014**, *8*, 62–68. [CrossRef]
221. Paździor, K.; Bilińska, L.; Ledakowicz, S. A review of the existing and emerging technologies in the combination of AOPs and biological processes in industrial textile wastewater treatment. *Chem. Eng. J.* **2019**, *376*, 120597. [CrossRef]
222. Szczuka, A.; Berglund-Brown, J.P.; Chen, H.K.; Quay, A.N.; Mitch, W.A. Evaluation of a Pilot Anaerobic Secondary Effluent for Potable Reuse: Impact of Different Disinfection Schemes on Organic Fouling of RO Membranes and DBP Formation. *Environ. Sci. Technol.* **2019**, *53*, 3166–3176. [CrossRef]
223. Miralles-Cuevas, S.; Darowna, D.; Wanag, A.; Mozia, S.; Malato, S.; Oller, I. Comparison of UV/H_2O_2, UV/$S_2O_8$2−, solar/Fe(II)/H_2O_2 and solar/Fe(II)/$S_2O_8$2− at pilot plant scale for the elimination of micro-contaminants in natural wat. *Chem. Eng. J.* **2017**, *310*, 514–524. [CrossRef]
224. Garcia-Segura, S.; Bellotindos, L.M.; Huang, Y.H.; Brillas, E.; Lu, M.C. Fluidized-bed Fenton process as alternative wastewater treatment technology—A review. *J. Taiwan Inst. Chem. Eng.* **2016**, *67*, 211–225. [CrossRef]

225. Hamza, R.A.; Iorhemen, O.T.; Tay, J.H. Occurrence, impacts and removal of emerging substances of concern from wastewater. *Environ. Technol. Innov.* **2016**, *5*, 161–175. [CrossRef]
226. De Araújo, K.S.; Antonelli, R.; Gaydeczka, B.; Granato, A.C.; Malpass, G.R.P.; Garcia-Segura, S.; Bellotindos, L.M.; Huang, Y.H.; Brillas, E.; Lu, M.C.; et al. Advanced oxidation processes for antibiotics removal: A review. *Chem. Eng. J.* **2016**, *46*, 211–225.
227. Michael, I.; Hapeshi, E.; Michael, C.; Varela, A.R.; Kyriakou, S.; Manaia, C.M.; Fatta-Kassinos, D. Solar photo-Fenton process on the abatement of antibiotics at a pilot scale: Degradation kinetics, ecotoxicity and phytotoxicity assessment and removal of antibiotic resistant enterococci. *Water Res.* **2012**, *46*, 5621–5634. [CrossRef]
228. Matsushita, T.; Honda, S.; Kuriyama, T.; Fujita, Y.; Kondo, T.; Matsui, Y.; Shirasaki, N.; Takanashi, H.; Kameya, T. Identification of mutagenic transformation products generated during oxidation of 3-methyl-4-nitrophenol solutions by orbitrap tandem mass spectrometry and quantitative structure–activity relationship analyses. *Water Res.* **2018**, *129*, 347–356. [CrossRef]
229. Tišler, T.; Pintar, A. *Evolution of Toxicity and Estrogenic Activity Throughout AOP's Surface and Drinking Water Treatment BT—Applications of Advanced Oxidation Processes (AOPs) in Drinking Water Treatment*; En; Gil, A., Galeano, L.A., Vicente, M.Á., Eds.; Springer International Publishing: Cham, Switzerland, 2019; pp. 387–403. ISBN 978-3-319-76882-3.
230. Arvai, A.; Jasim, S.; Biswas, N. Bromate Formation in Ozone and Advanced Oxidation Processes. *Ozone Sci. Eng.* **2012**, *34*, 325–333. [CrossRef]
231. Bourgin, M.; Beck, B.; Boehler, M.; Borowska, E.; Fleiner, J.; Salhi, E.; Teichler, R.; von Gunten, U.; Siegrist, H.; McArdell, C.S. Evaluation of a full-scale wastewater treatment plant upgraded with ozonation and biological post-treatments: Abatement of micropollutants, formation of transformation products and oxidation by-products. *Water Res.* **2018**, *129*, 486–498. [CrossRef]
232. Bourgin, M.; Borowska, E.; Helbing, J.; Hollender, J.; Kaiser, H.-P.; Kienle, C.; McArdell, C.S.; Simon, E.; von Gunten, U. Effect of operational and water quality parameters on conventional ozonation and the advanced oxidation process O_3/H_2O_2: Kinetics of micropollutant abatement, transformation product and bromate formation in a surface water. *Water Res.* **2017**, *122*, 234–245. [CrossRef]
233. Semitsoglou-Tsiapou, S.; Templeton, M.R.; Graham, N.J.D.; Mandal, S.; Hernández Leal, L.; Kruithof, J.C. Potential formation of mutagenicity by low pressure-UV/H_2O_2 during the treatment of nitrate-rich source waters. *Environ. Sci. Water Res. Technol.* **2018**, *4*, 1252–1261. [CrossRef]
234. Sharma, A.; Ahmad, J.; Flora, S.J.S. Application of advanced oxidation processes and toxicity assessment of transformation products. *Environ. Res.* **2018**, *167*, 223–233. [CrossRef]
235. Lai, W.W.P.; Hsu, M.H.; Lin, A.Y.C. The role of bicarbonate anions in methotrexate degradation via UV/TiO_2: Mechanisms, reactivity and increased toxicity. *Water Res.* **2017**, *112*, 157–166. [CrossRef]
236. Magdeburg, A.; Stalter, D.; Oehlmann, J. Whole effluent toxicity assessment at a wastewater treatment plant upgraded with a full-scale post-ozonation using aquatic key species. *Chemosphere* **2012**, *88*, 1008–1014. [CrossRef]
237. Li, X.; Shi, H.; Li, K.; Zhang, L. Combined process of biofiltration and ozone oxidation as an advanced treatment process for wastewater reuse. *Front. Environ. Sci. Eng.* **2015**, *9*, 1076–1083. [CrossRef]
238. Dantas, R.F.; Rossiter, O.; Teixeira, A.K.R.; Simões, A.S.M.; Da Silva, V.L. Direct UV photolysis of propranolol and metronidazole in aqueous solution. *Chem. Eng. J.* **2010**, *158*, 143–147. [CrossRef]
239. Chan, S.H.S.; Wu, T.Y.; Juan, J.C.; Teh, C.Y. Recent developments of metal oxide semiconductors as photocatalysts in advanced oxidation processes (AOPs) for treatment of dye waste-water. *J. Chem. Technol. Biotechnol.* **2011**, *86*, 1130–1158. [CrossRef]
240. Lv, Y.; Yu, L.; Huang, H.; Liu, H.; Feng, Y. Preparation, characterization of P-doped TiO_2 nanoparticles and their excellent photocatalystic properties under the solar light irradiation. *J. Alloys Compd.* **2009**, *488*, 314–319. [CrossRef]
241. Hamadanian, M.; Reisi-Vanani, A.; Majedi, A. Synthesis, characterization and effect of calcination temperature on phase transformation and photocatalytic activity of Cu, S-codoped TiO_2 nanoparticles. *Appl. Surf. Sci.* **2010**, *256*, 1837–1844. [CrossRef]
242. Janus, M.; Choina, J.; Morawski, A.W. Azo dyes decomposition on new nitrogen-modified anatase TiO_2 with high adsorptivity. *J. Hazard. Mater.* **2009**, *166*, 1–5. [CrossRef]

243. Liu, Y.; Zeng, X.; Hu, X.; Hu, J.; Zhang, X. Two-dimensional nanomaterials for photocatalytic water disinfection: Recent progress and future challenges. *J. Chem. Technol. Biotechnol.* **2019**, *94*, 22–37. [CrossRef]
244. Lai, C.; Wang, M.M.; Zeng, G.M.; Liu, Y.G.; Huang, D.L.; Zhang, C.; Wang, R.Z.; Xu, P.; Cheng, M.; Huang, C.; et al. Synthesis of surface molecular imprinted TiO_2/graphene photocatalyst and its highly efficient photocatalytic degradation of target pollutant under visible light irradiation. *Appl. Surf. Sci.* **2016**, *390*, 368–376. [CrossRef]
245. Tisa, F.; Abdul Raman, A.A.; Wan Daud, W.M.A. Applicability of fluidized bed reactor in recalcitrant compound degradation through advanced oxidation processes: A review. *J. Environ. Manag.* **2014**, *146*, 260–275. [CrossRef]
246. Briones, R.M.; De Luna, M.D.G.; Lu, M.C. Optimization of acetaminophen degradation by fluidized-bed Fenton process. *Desalin. Water Treat.* **2012**, *45*, 100–111. [CrossRef]
247. Anotai, J.; Sakulkittimasak, P.; Boonrattanakij, N.; Lu, M.C. Kinetics of nitrobenzene oxidation and iron crystallization in fluidized-bed Fenton process. *J. Hazard. Mater.* **2009**, *165*, 874–880. [CrossRef] [PubMed]

Article

Transportation of Different Therapeutic Classes of Pharmaceuticals to the Surface Water, Sewage Treatment Plant, and Hospital Samples, Malaysia

Fouad Fadhil Al-Qaim [1,2,*], Zainab Haider Mussa [1], Ali Yuzir [1], Nurfaizah Abu Tahrim [3], Norbaya Hashim [1,4] and Shamila Azman [5]

1 Malaysia-Japan International Institute of Technology (MJIIT), Universiti Teknologi Malaysia (UTM), 54100 Kuala Lumpur, Malaysia; zp69014@yahoo.com (Z.H.M.); muhdaliyuzir@utm.my (A.Y.); norbaya@nahrim.gov.my (N.H.)

2 Department of Chemistry, College of Science for Women, University of Babylon, P.O. Box 4, 51 Babylon, Iraq

3 Centre for Water Research and Analysis (ALIR), Faculty of Science and Technology, Universiti Kebangsaan Malaysia, 43600 Bangi, Selangor, Malaysia; nfaizah@ukm.edu.my

4 National Hydraulic Research Institute of Malaysia, Lot 5377, Jalan Putra Permai, 43300 Seri Kembangan, Selangor, Malaysia

5 Department of Environmental Engineering, Faculty of Civil Engineering, Universiti Teknologi Malaysia, 81310 Skudai Johor, Malaysia; shamila@utm.my

* Correspondence: fouadalkaim@yahoo.com; Tel.: +60-173394821

Received: 27 April 2018; Accepted: 6 June 2018; Published: 11 July 2018

Abstract: All pharmaceuticals are separated chromatographically using the liquid chromatography-time of flight/mass spectrometry (LC-ToF/MS) on a 5 µm, 2.1 mm × 250 mm, C18 column at 0.3 mL/min. The recovery is investigated at two spiking levels, 10 and 1 ng/mL; the mean recovery is higher than 77, 84, and 93% in sewage treatment plants (STP) influent, STP effluent, and surface water, respectively. The limit of quantification (LOQ) averages 29, 16, 7, and 2 ng/L in STP influent, STP effluent, surface water, and drinking water, respectively. The matrix effect is also evaluated in STP influent and effluent. It is observed that sulfamethoxazole, prednisolone, ketoprofen, and glibenclamide are highly impacted compared to other compounds, −99, −110, 77, and 91%, respectively. The results show that six out of nine pharmaceuticals, namely atenolol, acetaminophen, theophylline, caffeine, metoprolol, and sulfamethoxazole are detected in STP influent, STP effluent, and surface water. However, the means of concentration are 561, 3305, 1805, 3900, 78, and 308 ng/L for atenolol, acetaminophen, theophylline, caffeine, metoprolol, and sulfamethoxazole, respectively, in STP influent. Caffeine and acetaminophen are detected with the highest concentration, reaching up to 8700 and 4919 ng/L, respectively, in STP influent.

Keywords: transportation of pharmaceuticals; Malaysian aquatic environment; pharmaceutical consumption; LC-ToF/MS

1. Introduction

It is well known that different therapeutic classes of pharmaceuticals are used for the treatment of some diseases in the human body since they are biologically active compounds used for this purpose. However, the occurrence of some pharmaceuticals in surface water may be due to the bodily excretion of metabolized and un-metabolized pharmaceutical compounds into septic wastewater, which is then discharged to surface water. Although these concentrations are very low (ng/L), they are a big concern for their potential impact on the aquatic environment [1]. In the environmental analysis of pharmaceuticals, many methods have been reported in literature using liquid chromatography (LC).

The reason why LC was used in the analysis of these pharmaceuticals is related to the low volatility and high hydrophilicity of most of the pharmaceuticals. A gas chromatography instrument has also been used for the analysis of pharmaceuticals [2,3]; however, the derivatization of pharmaceuticals is needed to attain volatility and stability prior to the injection into the gas chromatography instrument [4]. The quantification analysis of pharmaceuticals is challenging due to high interference with other organic pollutants and low concentrations present in real samples [5]. Generally, a sample preparation method is required prior to instrumental analysis. A liquid-phase micro-extraction is a relatively newly developed extraction process consuming low solvent consumption [6].

This procedure requires small volume, and it is more suitable for biological samples such as blood and urine. [7]. So far, solid phase extraction is considered one of the most frequently employed extraction techniques in the analysis of pharmaceuticals in water, as it offers high selectivity, precision, and extraction efficiency [8–10]. The most common solid phase extraction (SPE) materials that allow the retention of a wide variety of compounds are the copolymer poly (divinylbenzene-co-*N*-vinylpyrrolidone) (Oasis HLB) that has both hydrophilic and lipophilic retention characteristics, and it can be used to retain both polar and non-polar compounds [10,11]. So far, very limited literature on the multi-residue analysis method for pharmaceuticals has been reported in Malaysia.

Therefore, the aim of this present study is to investigate the possibility of analyzing different therapeutic classes of pharmaceuticals in different bodies of water using a single solid phase extraction method by developing a very accurate and selective liquid chromatography-time of flight/mass spectrometry (LC-ToF/MS) method.

Hence, this study was conducted to develop a sensitive and accurate method for the determination of nine pharmaceuticals which are selected based on the national consumption report in Malaysia [12].

The aim of this work is to develop and validate a comprehensive analytical LC-ToF/MS method that can simultaneously detect and quantify a wide spectrum of pharmaceuticals in water samples. One single extraction method is applied to investigate and quantify the studied pharmaceuticals in surface water, sewage treatment plant (STP) influent, STP effluent, and hospital effluent.

2. Materials and Methods

2.1. Consumption of the Pharmaceuticals

The general description of the studied pharmaceutical compounds was overviewed [13]. Atenolol and metoprolol are called beta blockers, which are used for the treatment of high and low blood pressure and to prevent heart attack. A non-prescription compound also known as paracetamol, Acetaminophen is commonly used for its analgesic and antipyretic effects; its therapeutic effects are similar to salicylates. The non-prescription pharmaceutical stimulant xanthine compounds caffeine and theophylline are included in this present study. Theophylline is used to relax the muscles in the airway, making breathing easier, while caffeine is responsible for the stimulation of the central nervous system in the body.

Sulfamethoxazole is an antibacterial used to reduce the impact of bacterial synthesis of dihydrofolic acid. Prednisolone is one of the steroid compounds used to help reduce the symptoms of asthma, such as wheezing in children. Ketoprofen is nonsteroidal anti-inflammatory drug, which is used for the symptomatic treatment of acute and chronic rheumatoid arthritis. Glibenclamide is called glyburide, an antidiabetic drug, which is used to reduce the blood glucose in patients with non-insulin-dependent diabetes mellitus (Type II diabetes). In Malaysia, the Ministry of Health annually publishes a statistical report on drug consumption. Table 1 presents the defined daily doses (DDD) of the studied pharmaceuticals per thousand inhabitants between 2011 and 2014 in Malaysia. The DDD values are based on the Anatomical Therapeutic Chemical (ATC) classification system by the World Health Organization (WHO) [14]. The annual consumption of these pharmaceuticals can be calculated using the following formula:

$$\text{Consumption (kg)} = \text{DDD(g)} \times \text{DDD/1000 inh} \times \text{Population/1,000,000} \times 366 \quad (1)$$

where DDD is the defined daily dose and DDD/1000 inh is the number of daily doses consumed per 1000 inhabitants in one year.

Table 1. Defined daily doses (DDD) and the consumption of the selected pharmaceuticals in Malaysia (MOH 2014).

Compound	DDD (mg) [a]	Consumption (kg/year)			
		2011	2012	2013	2014
Atenolol	75 (O, P)	8184	8233	8428	9094
Acetaminophen	3000 (O, P, R)	249,358	255,444	268,814	272,690
Theophylline	400 (O, R, P)	-	-	-	-
Caffeine	400 (O, P)	-	-	-	-
Metoprolol	150 (O, P)	13,025	14,387	15,393	15,689
Prednisolone	10 (O)	736	649	589	479
Glibenclamide (O)	10 (O)	392	616	458	435
Sulfamethoxazole [b]	2000 (O)	-	-	-	-
Ketoprofen [b]	NA	-	-	-	-
Population (10^7 inhabitants)		2.9062	2.9510	2.9915	3.0261

[a] WHO (2018), [b] Means the compound not listed as top 50 pharmaceuticals consumed in Malaysia. O = Oral, P = Parenteral, R = Rectal, NA: not available.

It was observed from Table 1 that acetaminophen has the highest consumption levels during the four years compared to the other pharmaceutical compounds. Furthermore, it could be considered an over-the-counter drug, and it is consumed in three different ways: orally, parenterally, and rectally.

All of the studied pharmaceuticals have been presented in Figure 1. Non-prescription pharmaceutical compounds, such caffeine and theophylline, were selected for their prevalence in very commonly consumed drinks, such as tea, coffee, milo, and Pepsi, and, furthermore, due to their frequent detection as reported in previous studies [11,15–19]. Acetaminophen was selected as a non-prescribed and/or prescribed pharmaceutical compound.

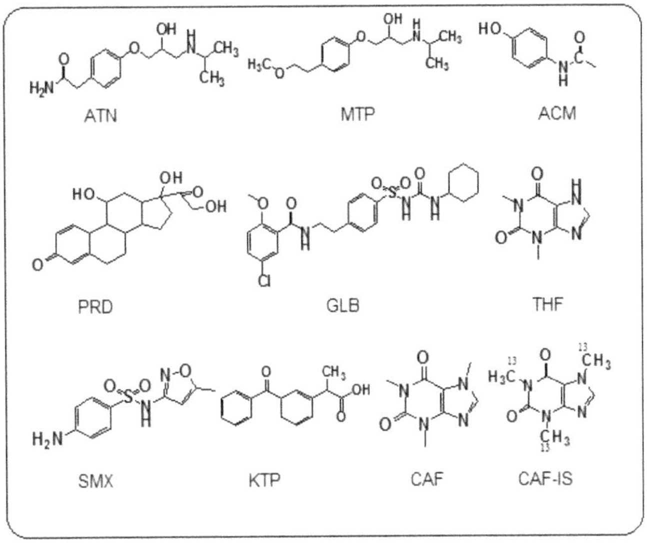

Figure 1. Chemical structures of the studied pharmaceuticals.

2.2. Reagents and Materials

All standards were pure with (≥98%) of atenolol, acetaminophen, theophylline, caffeine, caffeine-$^{13}C_3$ (internal standard, IS), metoprolol, sulfamethoxazole, prednisolone, ketoprofen, and glibenclamide, which were obtained from Sigma-Aldrich (St. Louis, MO, USA). Deionized water (DIW) was collected from the water analysis and research lab at Universiti Kebangsaan Malaysia. HPLC-grade methanol (MeOH), acetonitrile (ACN), methyl tertiary butyl ether (MTBE), acetone, and formic acid (FA) were supplied by Merck (Darmstadt, Germany). The cartridges used for SPE were Oasis HLB (3cc, Waters, Milford, MA, USA).

2.3. Sample Collection

All samples were collected from Nilai and Seremban, Malaysia, and then shipped to the laboratory on the same day. Eight points, as shown in the map (see Figure 2), were chosen to study the fate of nine pharmaceutical compounds. Samples were collected from four STPs (STP1, STP2, STP3, and STP4) and two hospitals (HSP1 and HSP2). Samples were also collected from the recipient rivers at two points (SW1 and SW2). The treatment process in all STPs was an oxidation ditch, while it was a rotating biological contractor in the hospitals. The frequency of sampling was for three months in 2014. Samples were collected on the same day, within three hours in the morning, at a fixed volume (1.0 L) for each point; the sampling interval was every month. One liter amber glass bottles were rinsed in the field twice before sample collection. A polyethylene plastic bucket was used to collect wastewater samples and fill the glass bottles. All safety was taken into account during sampling. The sampler used disposable gloves to prevent any contamination by the personal care products from the sample. A plastic bucket was used to collect the samples. All samples were filtered by 0.7 µm GF/F filter (Whatman, Little Chalfont, UK) to remove any solid matter suspended in the samples. All filtered samples were kept at 4 °C until the solid phase extraction experiments.

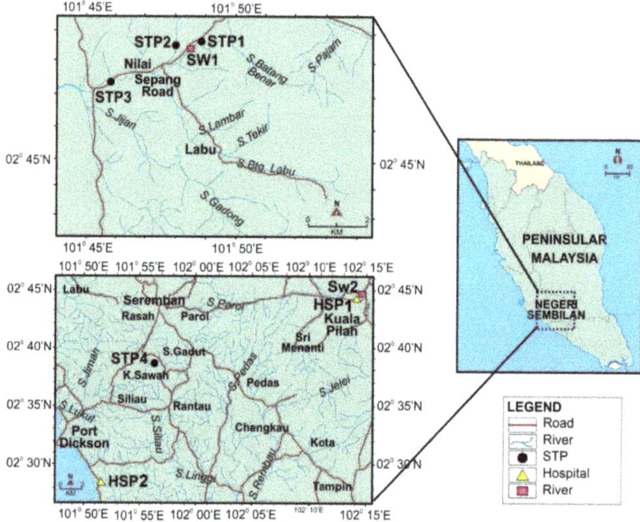

Figure 2. Map to describe sampling points.

2.4. Instrumental and Extraction Method

Separation of pharmaceuticals was performed on the liquid chromatography (LC) instrument (Dionex, Sunnyvale, CA, USA). 30 µL of sample was injected at 0.3 mL/min. All selected pharmaceuticals were analyzed in the positive mode (PI). Two mobile phases were used; (A) 0.1% FA in DIW and (B) ACN-MeOH (3:1, *v*/*v*) at 0.3 mL/min. The gradient elution is as follow:

5% B (0 min) → 60% B (linear increased in 3 min) → 97% B (linear increased in 3 min) → 97% B (hold 5 min) → 5% B (linear decreased in 0.1 min) → 5% B (hold 5 min).

All analytes were acquired using an independent reference spray via the LockSpray interference to ensure accuracy and reproducibility (mass spectrometry (MS) capillary voltages, 4000 (PI); drying gas flow rate, 8.0 L/min; drying gas temperature, 190 °C; and nebulizer pressure, 4.0 bar). A mixture of sodium hydroxide and FA was used as the lock mass *m/z* 90.9766 to 974.8132. Accurate masses were calculated using the software Daltons Data Analysis incorporated in the instrument. Samples of 500, 250, 100 mL from surface water, sewage treatment plant and hospital effluent, and sewage treatment plant influent, respectively, were filtered by 0.7 μm GF/F filter (Whatman, UK) to remove any solid matter suspended in the samples. All filtered samples were kept at 4 °C until solid phase extraction. The sample extraction method was provided using Oasis hydrophilic-lypophilic balanced (HLB) (3 cc, 60 mg) cartridges (Waters, Milford, MA, USA). To achieve all SPE experiments, a vacuum manifold was used for this extraction procedure. SPE cartridges were pre-conditioned with 2 mL of methanol and 2 mL of deionized water (DIW) before sample loading. Water samples were loaded at a flow rate of 3 mL/min under vacuum. To exclude water residue from the cartridge, it was dried under vacuum for 15 min. After that, analytes were eluted by passing 5 mL of MTBE and 5 mL of (MeOH-ACN, 50:50, *v*/*v*). Then, eluents were dried by flowing a stream of nitrogen gas. A 0.5 mL of solvent was added to reconstitute the extracted analytes, which were filtered by 0.45 μm (Nylon syringe) before injection. Each experiment was repeated three times to find the precision of the injection using LC-ToF/MS. Individual stock standard solutions and caffeine $^{13}C_3$ as an internal standard solution (1000 μg/mL) were prepared in MeOH by dissolving 0.01 g of compound in 10 mL of methanol. Stock solutions were kept at −20 °C until further experiments. Working solutions were prepared by serial dilutions of stock standard solution with MeOH-DIW (1:9, *v*/*v*) solvent.

2.5. Method Validation

Selectivity is the ability of an analytical method to differentiate and quantify the targeted compounds in the presence of other sample components [20]. The method selectivity was investigated by analyzing a blank solvent sample MeOH:DIW (10:90, *v*/*v*), an effluent STP sample spiked with nine pharmaceuticals and one IS, and an effluent STP un-spiked sample. It was observed that the LC-ToF/MS method has a good ability to select the target compounds from different samples. To assess intra-day precision, three concentrations (8, 40 and 200 ng/mL) of mixture compounds were injected three times using liquid chromatography–time of flight/mass spectrometry. To assess inter-day precision, samples were analyzed with the same above concentrations on three separate days. Five replicates (n = 5) were performed on the same day (intra-day precision) and at different days (inter-day precision). Recovery was investigated in different samples: surface water, the influent of a sewage treatment plant, and the effluent of a sewage treatment plant. Standard solution mixtures of 1 and 10 ng/mL (n = 5) were spiked in the samples and extracted using solid phase extraction. The recoveries were evaluated by comparing the peak area of the extracted samples to the peak area of the standard solutions. The recoveries (R%) were calculated based on the following formula:

$$R\% = \frac{(A_{SP} - A_{UN})}{A_S} \times 100\% \quad (2)$$

where A_{SP} is the peak area of a compound in an extract, A_{UN} is the peak area of a compound in a sample, and A_s is the peak area of a compound in the standard solution.

Linearity was investigated by generating the calibration curve for each analyte. Four to five points of calibration curves were generated by injecting mixture solutions prepared from the standard stock solution. Concentrations used to create the calibration curves ranged from each analyte's instrumental quantification limit (IQL) up to 400 ng/mL. Calibration curves were generated for each compound by plotting the peak area against the concentration of each compound using the linear regression model. The determination coefficient, $R^2 \geq 0.993$, was obtained for all analytes. The instrumental quantification limit (IQL) was the lowest concentration corresponding to the signal-to-noise ratio (S/N) ratio ≥ 10. The limit of quantification (LOQ) for the whole method in the different matrices was estimated using the following formula [21]:

$$\text{LOQ} = \frac{\text{IQL} \times 100}{\text{R\%} \times \text{CF}} \quad (3)$$

where IQL is the instrumental quantification limit (ng/L), R (%) is the recovery of the compound, and CF is the concentration factor which corresponds to 2000, 1000, 500, and 200 for drinking water, surface water, STP and HSP effluent, and STP influent, respectively. The identification and quantification of pharmaceutical compounds was based on retention times (Rt) and mass value (*m/z*) for each analyte. A 0.02 Da narrow window was applied for all analytes to be extracted and quantified in real samples.

2.6. Matrix Effects

The matrix effect (ME%) was evaluated based on the signal intensity of the analytes in a sample. However, it was calculated according to this procedure: Sample extracts of STP influent and effluent were spiked at a level of 10 ng/mL of pharmaceuticals (n = 3), and then it could be injected to LC–ToF/MS. The following formula was used to estimate the matrix effect.

$$\text{ME\%} = \frac{A_S - (A_{SP} - A_{UN})}{A_S} \times 100\% \quad (4)$$

where A_S is the peak area of the compound in the standard solution, A_{SP} is the peak area of the compound in the extract of the sample, and A_{UN} is the peak area of the compound in the un-spiked extract.

3. Results and Discussion

Atenolol, metoprolol, sulfamethoxazole, prednisolone, ketoprofen, and glibenclamide were selected as the top prescribed pharmaceuticals in Malaysia [12]. The most commonly used non-prescription drugs were acetaminophen, theophylline, and caffeine. An example of chromatographical separation (Figure 3) was provided by using a gradient elution program as described in the previous section. The intensity of the pharmaceutical compounds varied strongly. This variation may be due to the diversity of physico-chemical properties among the selected pharmaceuticals under electrospray ionization conditions. However, the LC-chromatogram was more sufficient for analysis of the studied pharmaceuticals at the expected ambient environmental concentrations.

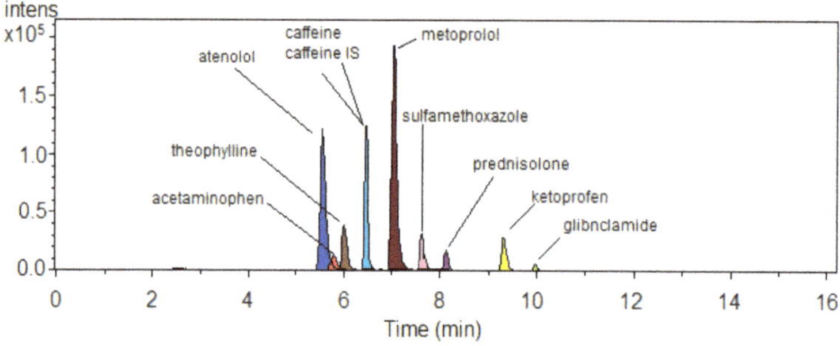

Figure 3. Representative liquid chromatography-time of flight/mass spectrometry of a standard solution of the pharmaceutical compounds determined in this study.

3.1. Elution Solvent Effect

The variety of physico-chemical properties of pharmaceutical compounds provided varying results among the elution solvents tested in the preliminary experiments. The best overall solvent elution recoveries were achieved using 5 mL of MTBE + 5 mL of methanol:acetone (50:50, *v*/*v*) without a pH adjustment of the sample. Recoveries for the compounds tested are shown in Figure 4. It is well known that extraction of pharmaceuticals from the Hydrophilic-Lipophilic Balance sorbent (HLB-oasis cartridge) could be impacted by the polarity and non-polarity of the solvent and depends on the type of analytes. Various elution solvents combining methanol, acetone, and methyl tertiary butyl ether have been tested in this study. These elution solvents include the following: Elu **A**: 10 mL of MeOH, Elu **B**: 10 mL of acetone, Elu **C**: 10 mL of methanol:acetone solution (50:50, *v*/*v*), Elu **D**: 10 mL of MTBE, and Elu **E**: 5 mL of MTBE + 5 mL of methanol:acetone (50:50, *v*/*v*). On average, the analytes were recovered by 84% with eluent **E**, 63% with eluent **D**, 67.5% with eluent **C**, 65.1% with eluent **B**, and 48% with eluent **A**.

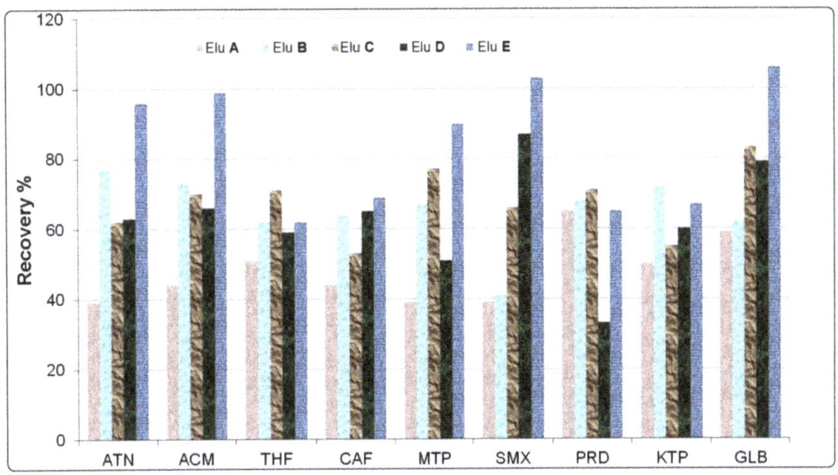

Figure 4. Influence of elution solvent on the recovery of the studied pharmaceuticals at 10 ng/mL of standards (n = 3).

Slightly-polar compounds, such as atenolol, acetaminophen, metoprolol, sulfamethoxazole, and glibenclamide, were recovered at less than 60% with methanol as the elution solvent (**A**). By reducing the polarity of the solvent, it was observed that most of the compounds were well recovered. A polar–nonpolar elution solvent, such as eluent **E**, was the best choice in this present study. Low recovery for slightly-polar pharmaceutical compounds may be attributed to the poor elution from the HLB sorbent or poor retention on the sorbent, whilst most of compounds were very well recovered (≥90%) using eluent **E** (5 mL of MTBE + 5 mL of methanol: acetone (50:50, *v*/*v*)). Recoveries of less than 50% were observed for the prednisolone and ketoprofen in the presence of eluent **E**. Theophylline and caffeine, in the same way, are believed to be poorly retained in the polymeric sorbent without pH adjustment. However, this low recovery is not an obstacle to quantify theophylline and caffeine in real samples, as they have a reliably low limit of quantification.

3.2. ToF Screening and Confirmation

Future strategies in LC-ToF/MS method development would include the use of electrospray ionization modes to enhance detection methodology. In addition, the development of good chromatographical separation using different mobile phases (data not shown) provides a highly sensitive and selective method to separate and quantify the compounds in real samples.

Two types of ionization, either positive or negative modes, were optimized. However, positive ionization (PI) was selected to monitor and quantify the analytes in the samples. It was found that, at the PI mode, the *S/N* ratio for all selected pharmaceuticals was the highest (data not shown). Two compounds, ketoprofen and glibenclamide, were also identified in the negative ionization (NI) mode, but their intensities were very low. Thus, all pharmaceutical compounds were analyzed in positive ionization mode.

The low detection limit and the possibility of interference with other organic pollutants in a real sample that has mass-to-charge value close to that of studied pharmaceuticals are one of the most challenging fields in quantitative analysis. In order to reduce this challenge and to increase the selectivity of ToF/MS measurements, a narrow, accurate mass interval was used to reconstruct the chromatographic traces levels. Extracted ion chromatograms (EIC) were typically extracted using a 0.02 Da for all studied pharmaceuticals. However, reducing the mass window resulted in an enhancement of the detection limit in influent and effluent sewage treatment plants and a complete loss of interferences from contaminants.

Petrovic et al. observed that reducing the mass window from 100 to 20 mDa resulted in an almost 15-fold increase of the signal-to-noise ratio and in an almost complete loss of the interferences from the isobaric contaminant ions for carbamazepine in urban wastewater [22].

In all cases, the accurate mass of the protonated $[M + H]^+$ molecular ions were applied for confirmation and quantification purposes. Accurate mass data for the molecular ions was processed through the software Brucker Daltons Data Analysis, which provided the elemental formula and mass errors. Figure 5 shows an example of atenolol analysis in the influent of sewage treatment plants using Brucker software. It could be observed that the elemental formula ($C_{14}H_{23}N_2O_3$) has -1.8 ppm, which is an accurate value to confirm that this formula belongs to atenolol. The other exact mass measurements, retention times, elemental composition, and mass errors were presented in Table 2.

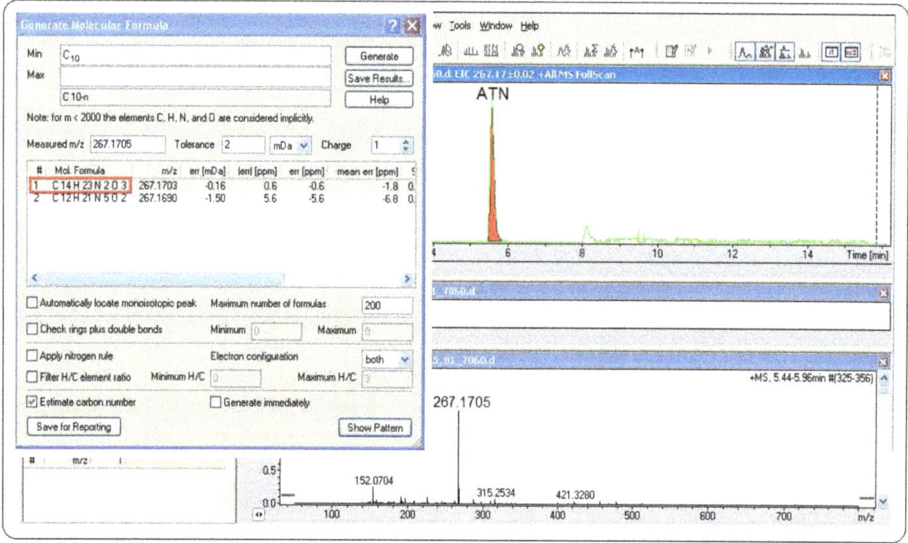

Figure 5. Bruker Daltons Data Analysis software with atenolol as an example.

Table 2. Retention times and accurate mass measurements of molecular ions of target pharmaceuticals in a standard solution.

Compound	Elemental Composition	Theoretical Mass m/z	Extracted Mass m/z	RT (min)	Collision Energy (eV)	Error	
						mDa	ppm
Atenolol	$[M + H]^+$ $C_{14}H_{23}N_2O_3$	267.1704	267.1700	5.55	10	−0.4	−1.5
Acetaminophen	$[M + H]^+$ $C_8H_{10}NO_2$	152.0712	152.0718	5.79	10	0.6	3.9
Theophylline	$[M + H]^+$ $C_7H_9N_4O_2$	181.0731	181.0728	5.99	10	−0.3	−1.7
Caffeine	$[M + H]^+$ $C_8H_{11}N_4O_2$	195.0921	195.0925	6.47	10	0.4	2.1
Metoprolol	$[M + H]^+$ $C_{15}H_{26}NO_3$	268.1910	268.1913	7.03	10	0.3	1.1
Sulfamethoxazole	$[M + H]^+$ $C_{10}H_{12}N_3O_3S$	254.0592	254.0604	7.64	10	1.2	4.7
Prednisolone	$[M + H]^+$ $C_{21}H_{29}O5$	361.2020	361.2019	8.11	10	−0.1	−0.3
Ketoprofen	$[M + H]^+$ $C_{16}H_{15}O_3$	255.1021	255.1015	9.32	10	−0.6	−2.4
Glibenclamide	$[M + H]^+$ $C_{23}H_{29}ClN_3O_5S$	494.1511	494.1522	9.95	10	1.1	2.2

3.3. Method Validation

The linearity of the external calibration curve method ranged from IQL to 400 ng/mL for all compounds. It was observed from Table 3 and Figure S1 that (4–5) points were generated to achieve correlation coefficients (R^2) ≥ 0.993 using linear regression. The IQL for each analyte was determined using pure standards that were analyzed using the LC–ToF/MS method. The IQL was determined to be the concentration with an *S/N* ratio ≥10. A wide range of IQLs were obtained, because they depend on the sensitivity of the instrument and the ionization efficiency of the analyte in an electrospray ionization (ESI) source. The IQLs for all pharmaceuticals ranged between 0.3 and 8 ng/mL except for prednisolone, which was 40 ng/mL. This high value is not an obstacle for developing the method since most pharmaceuticals have very good IQLs compared to prednisolone. The LOQs over the entire method were calculated using Equation (1), in which the concentration factors and matrix effects of different environmental samples were considered. In drinking water (DW), the LOQ ranged between 0.3 and 8.2 ng/L. In effluent of STP, the LOQ ranged between 6.5 and 50.3 ng/L, whereas the LOQ ranged between 11.1 and 83 ng/L in influent of STP. The findings in this present study were in agreement with our previous studies [11,19].

Table 3. Method validation parameters.

Compound	Equation (5 Points)	R^2	Range ng/mL	IQL ng/mL	LOQ (ng/L)			
					INF STP	EFF STP	SW	DW
Atenolol	y = 2100x + 6486	0.9998	1.6–400	1.6	11.1	8.8	3.5	0.3
Acetaminophen	y = 350x − 118	0.9931	8–400	8	58.8	25.6	14.7	8.2
Theophylline	y = 664x − 103	0.9972	8–400	8	19.2	7.7	5.7	0.8
Caffeine	y = 663x + 8328	0.9963	4–400	4	22	17	8.4	5
Metoprolol	y = 2015x + 1938	0.9943	0.3–400	0.3	18.2	14.6	6.2	1.7
Sulfamethoxazole	y = 546x − 2963	0.9963	4–400	4	12.5	8.2	7.9	0.5
Prednisolone	y = 275x − 12888	0.9961	40–400	40	83	50.3	3.2	0.6
Ketoprofen	y = 449x − 3285	0.9950	8–400	8	14.3	6.5	7.7	1.4
Glibenclamide	y = 116x + 1389	0.9991	1.6–400	1.6	20.5	7.9	5.1	0.4

The precision of the method was evaluated based on the results of the analysis of three concentrations (8, 40, and 200 ng/mL) with three replications for each one on the same day and the results from inter-day precision from the other three different days. The values were compared with the standards; thus, all values demonstrated good results with RSD% ≤ 6.7% for intra-day precision and 11.7% for inter-day precision (see Table 4). Recoveries of the solid phase extraction method were compared to the recoveries from drinking water, surface water, and STP influent and effluent samples. Five samples of surface water, STP influent, and STP effluent were spiked at 1 and 10 ng/mL, and then extracted using HLB sorbent and eluent **E**. These set spikes are used to evaluate method performance over different matrices. CAF-$^{13}C_3$, as an internal standard, was spiked at the same concentrations to evaluate the relative recovery for caffeine as one of the selected compounds in this study. The absolute and relative recoveries are presented in Table 5. Recoveries for SPE trials, extracted and analyzed in triplicate, ranged from 30.7 to 79.6% in STP influent, 37.4 to 82.4% in STP effluent, and 41.2 to 86.4% in surface water at 1 ng/mL spiking level with a mean of 54, 61, and 69%, respectively. For 10 ng/mL spiking level, recoveries were better and ranged from 50.1 to 100.6% in STP influent, 61.2 to 106.9% in STP effluent, and 66.8 to 109.6% in surface water.

Table 4. Intra-day and inter-day precision and accuracy for all studied pharmaceuticals.

Compound	Concentration ng/mL	Intra-Day Precision (n = 5)			Inter-Day Precision (n = 5)		
		Found	RSD %	Accuracy %	Found	RSD %	Accuracy %
Atenolol	8	7.2	13.4	89.5	7.1	15.9	88.8
	40	37.4	11.3	93.4	41.2	12.4	103.1
	200	216	5.3	107.8	187.8	6.1	93.9
Acetaminophen	8	7.6	9.8	95.2	8.1	10.6	101.3
	40	36.2	7.9	90.4	37.8	9.1	94.6
	200	198.8	3.1	99.4	192.2	4.9	96.1
Theophylline	8	8.1	11.1	101.6	7.5	13.8	94.3
	40	37.1	9.7	92.8	35.2	11.2	87.9
	200	193.4	4.2	96.7	208.8	5.8	104.4
Caffeine	8	7.9	12.2	98.5	7.6	15.3	94.9
	40	37.6	10.5	93.9	36.7	12.1	91.7
	200	198.2	5.8	99.1	193.4	7.9	96.7
Metoprolol	8	7.5	11.5	93.2	8.2	13.9	102.7
	40	43.9	9.1	109.7	37.9	12.5	94.7
	200	195.6	4.7	97.8	201.4	5.8	100.7
Sulfamethoxazole	8	8.4	10.8	104.5	7.6	11.4	95.5
	40	35.7	7.6	89.3	39.6	8.3	98.9
	200	194.8	2.2	97.4	207.6	3.6	103.8
Prednisolone	8	7.4	9.6	92.9	8.1	10.4	100.6
	40	4.1	7.2	103.3	37.6	7.9	93.9
	200	201.4	2.9	100.7	210.6	4.8	105.3
Ketoprofen	8	7.2	8.1	90.1	7.1	9.6	88.3
	40	38.5	5.1	96.3	43.7	6.7	109.2
	200	199.6	1.4	99.8	220.2	3.1	110.1
Glibenclamide	40	34.9	14.3	87.2	42.5	15.7	106.3
	200	208.4	4.2	104.2	195.6	5.8	97.8

Table 5. Recovery for all studied pharmaceuticals at different spike levels 1 and 10 ng/mL.

Compound	Spike Level 10 ng/mL R% ± SD, n = 5			Spike Level 1 ng/mL R% ± SD, n = 5		
	INF STP	EFF STP	SW	INF STP	EFF STP	SW
Atenolol	88.6 ± 5.2	96.4 ± 7.8	104.2 ± 11.5	69.4 ± 5.3	78.3 ± 6.2	83.2 ± 7.5
Acetaminophen	92.6 ± 6.2	97.6 ± 7.5	99.8 ± 9.2	79.6 ± 5.3	81.4 ± 8.3	87.6 ± 9.3
Theophylline	50.1 ± 7.7	61.2 ± 10.4	70.2 ± 10.4	43.3 ± 3.8	43.2 ± 3.9	46.4 ± 5.4
Caffeine	56.7 ± 4.3	61.3 ± 4.7	66.8 ± 4.7	44.6 ± 7.7	50.2 ± 7.6	60.6 ± 5.5
	93.7 ± 6.8 [a]	*97.4 ± 3.5*	*103.7 ± 7.2*	*99.3 ± 3.1*	*102.2 ± 8.2*	*108.4 ± 6.9*
Metoprolol	86.4 ± 4.3	90.2 ± 4.7	93.4 ± 4.1	43.1 ± 5.2	59.2 ± 4.9	72.6 ± 4.8
Sulfamethoxazole	99.4 ± 9.62	102.6 ± 6.5	103.2 ± 10.6	58.2 ± 6.8	67.4 ± 5.1	78.2 ± 5.3
Prednisolone	63.6 ± 5.88	65.2 ± 5.2	71.4 ± 1.7	30.7 ± 5.8	37.4 ± 6.1	41.2 ± 9.2
Ketoprofen	57.4 ± 5.8	67.3 ± 4.8	73.4 ± 4.2	40.6 ± 3.1	46.4 ± 7.3	62.8 ± 5.9
Glibenclamide	100.6 ± 5.3	106.9 ± 10	109.6 ± 11.5	78.4 ± 6.1	82.4 ± 7.1	86.4 ± 10.3

[a] relative recovery (RR%). It was calculated using Caffeine-$^{13}C_3$.

The means of the recoveries were 77, 83, and 88% for STP influent, STP effluent, and surface water, respectively. Lower recoveries for the prednisolone and ketoprofen ranged from 30.7 to 41.2% and 40.6 to 48.8%, respectively, at 1 ng/mL in all samples; these are likely attributable to the unsuitability elution with eluent E as a non-polar to polar solvent. Although prednisolone and ketoprofen exhibited low recoveries, other pharmaceutical compounds were recovered well in this extraction method. In comparison to other previous studies, the recovery results were not considered surprising compared to those pharmaceutical compounds that recovered between 10 and 15% in wastewater samples, as reported by Ferrer et al. [23]. In the same way, Shaaban et al. reported that few compounds were recovered between 12.7 and 32.2% at a 100 μg/L spiking level [24]. Thus, this method could be acceptable for extracting all nine pharmaceuticals using a single solid phase extraction cartridge. All pharmaceuticals were eluted within 16.1 min, with atenolol as the first elute and glibenclamide as the last. A perfect chromatogram of nine pharmaceuticals and one internal standard spike in STP effluent is presented in Figure 6.

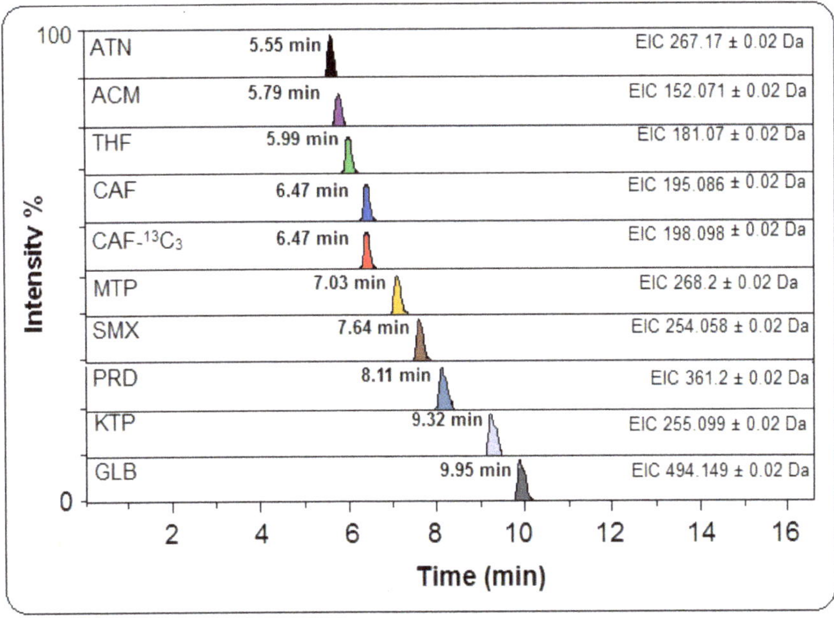

Figure 6. Extracted ion chromatogram (EIC) for all nine pharmaceuticals including caffeine $^{13}C_3$ as internal standard; spiking level 10 ng/mL in STP effluent.

3.4. Matrix Effect

As the developed sample preparation procedure involves an extraction process, ion suppression or enhancement is assessed through spiking the extracted influent and effluent of STP before injection to LC-ToF/MS at 10 ng/mL (see Figure 7). The matrix effect was evaluated as an enhancement or suppression according to Equation (3). Some pharmaceuticals at higher portions of acetonitrile (Rt > 7 min) were affected, in which signal suppression was 77 and 91% for ketoprofen and glibenclamide in influent of STP, respectively. Signal enhancement was also observed to be found at −99 and −110% for sulfamethxazole and prednisolone, respectively. These results indicate that the organic pollutants present in a matrix that elute at higher proportions of acetonitrile could suppress and/or enhance the ionization of the pharmaceutical compounds eluting at retention times longer than 7 min. The reason is related to the fact that at this time the polarity of elution mobile phase increases according to the elution program. The same phenomenon has been previously reported by Hernando et al. [25]. The matrix effect is highly dependent on the chromatographic gradient elution and the composition of the mobile phase; however, it was reported that some pharmaceuticals eluting at the beginning of the LC gradient were more heavily affected by the matrix effect as well [25].

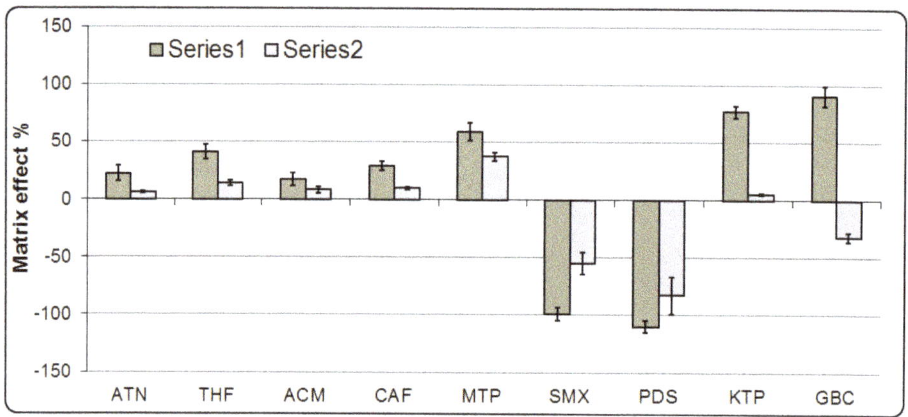

Figure 7. Percent matrix effect of pharmaceuticals (Spiking level 10 ng/mL, n = 5) from influent and effluent sewage treatment plant.

3.5. Occurrence of Pharmaceuticals in Water Samples

The LC-ToF/MS method described herein has been applied to samples collected from surface water, STP influent and effluent, and HSP effluent in Malaysia. The pharmaceuticals detected frequently by this method were six compounds (see Table 6), and some of them were presented in Figure S2. The most frequently detected compounds were non-prescription compounds, such as acetaminophen (75%), theophylline (100%), and caffeine (83.3%). Mean concentrations measured were 74 ng/L for acetaminophen, 38 ng/L for theophylline, and 540 ng/L for caffeine; the highest concentrations for these compounds were 110, 60, and 821 ng/L, respectively, in surface water. For the influent of sewage treatment, the mean concentrations measured were 3305, 1805, and 3900 ng/L for acetaminophen, theophylline, and caffeine, respectively; the highest concentrations for these compounds were 4919, 2722, and 8700 ng/L, respectively. All non-prescription pharmaceutical compounds were detected in all samples (100%).

For the effluent of sewage treatment plants, only theophylline was completely frequently detected (100%). Acetaminophen and caffeine were detected at 75% and 50%, respectively. The highest mean concentration measured, 360 ng/L, was for caffeine. The maximum concentrations for acetaminophen, theophylline, and caffeine were 122, 108, and 1190 ng/L, respectively. The frequency of detection was 100% for theophylline and caffeine and 50% for acetaminophen in hospital effluent, with the highest maximum concentrations of 3314, 628, and 2860 ng/L, respectively. Actually, the frequent detection of caffeine and theophylline in water samples is not surprising, as it is widely available in many drinks such as tea, coffee, cocoa, sport drinks, and soft drinks. Al-Qaim et al. reported that caffeine was detected in different beverages and tea drinks in Malaysia [26]. Pedrouzo et al. and Huggett et al. reported that caffeine was detected at a maximum concentration, reaching up to 9945 and 1056 ng/L, in wastewater and surface water, respectively [27,28]. In the same way, acetaminophen (prescribed and/or non-prescribed drug) was detected at a high concentration as well in all water samples. This finding may be due to its high levels of consumption by people as a therapeutic drug. However, the mean consumption of acetaminophen from 2011 to 2014 was 261,577 kg/y.

Table 6. Concentration of detected pharmaceuticals from surface water, STP influent and effluent, and HSP effluent.

Comp.	Influent STP/4 Points, 3 Replications (n = 12)			Effluent STP/4 Points, 3 Replications (n = 12)			Effluent Hospital/2 Points, 3 Replications (n = 6)			Surface Water/2 Points, 3 Replications (n = 6)		
	Frequency Detection	Mean (ng/L)	Range (ng/L)	Frequency Detection	Mean (ng/L)	Range (ng/L)	Frequency Detection	Mean (ng/L)	Range (ng/L)	Frequency Detection	Mean (ng/L)	Range (ng/L)
ATN	12:12	561	152–1009	12:12	89	20–181	6:6	216	61–485	6:6	35	19–55
ACM	12:12	3305	1891–4919	9:12	96	ND–122	3:6	1938	ND–3314	3:6	74	ND–110
THF	12:12	1805	902–2722	12:12	82	55–108	6:6	204	33–628	6:6	38	19–60
CAF	12:12	3900	980–8700	6:12	360	ND–1190	6:6	1600	73–2860	6:6	540	91–821
MTP	12:12	78	11–153	9:12	23	ND–36	6:6	221	44–606	6:6	124	34–190
SMX	9:12	308	ND–650	9:12	39	ND–52	3:6	147	ND–333	3:6	62	ND–118

Six prescribed pharmaceutical compounds were less frequently present in water samples. Two prescribed pharmaceuticals, prednisolone and glibenclamide, were not detected in all samples, while they are considered the top consumed pharmaceuticals in Malaysia. The reason may be attributed to their low consumption of 613 and 475 kg/y within four years (2011–2014). Ketoprofen and sulfamethoxazole were not listed as the top consumed pharmaceutical compounds from 2011 to 2014. Ketoprofen was also not detected in all samples. Although sulfamethoxazole was not listed as a top consumed compound in Malaysia, it was detected with 75% in STP influent and effluent. The mean concentration for sulfamethoxazole ranged between 52 and 650 ng/L in STP influent and effluent. Atenolol and metoprolol were the most prescribed pharmaceuticals present in water samples, and both compounds are the top consumed pharmaceutical in Malaysia. Atenolol was consumed by the human body, averaging 8485 kg/y within four years; it was frequently detected in STP influent and effluent, hospital effluent, and surface water. The highest concentration detected for atenolol was 1009 ng/L in STP influent. The mean concentration ranged from 35 to 561 ng/L.

The concentration of atenolol was 273 ng/L in the River Taff sample and 2702 ng/L in the wastewater effluent sample in the UK [29].

Metoprolol was frequently detected at 100%, 75%, 100%, and 100% in STP influent, STP effluent, HSP effluent, and surface water, respectively. The highest concentration detected for metoprolol was 606 ng/L in HSP effluent followed by 190 ng/L in surface water, 153 ng/L in STP influent, and 36 ng/L in STP effluent; however, these results were in line with the previous study [30].

4. Conclusions

The analysis and determination of pharmaceutical compounds within sewage treatment plant influent and effluent, hospital effluent, and surface water using SPE and LC-ToF/MS has been observed to be practical and effective. The prescription and non-prescription pharmaceuticals most likely found in Malaysian wastewater and surface waters were investigated and determined. The method performance presented indicates that the SPE and LC-ToF/MS techniques applied to routine analysis of sewage treatment plant influent and effluent, hospital effluent, and surface water for pharmaceuticals is sensitive and accurate for the majority of compounds tested, with detection limits averaging 29, 16, 7, and 2 ng/L in STP influent and effluent, surface water, and drinking water, respectively.

Re-constituted concentration and sample preparation were achieved by a solid phase extraction method after optimization of elution solvent. All studied pharmaceuticals were analyzed in the positive ionization mode, and they were separated in 16.1 min. Caffeine-$^{13}C_3$ was applied as an internal standard to investigate the method extraction efficiency; however, recovery was quite acceptable, wherein the means of the most analyzed pharmaceuticals ranged 50.1 to 100%, 61.2 to 106.9%, and 70.2 to 109.9% in STP influent, STP effluent, and surface water, respectively. The matrix effect was high for most of the compounds, especially those eluted after 7 min.

The results showed that six out of nine pharmaceuticals, namely atenolol, acetaminophen, theophylline, caffeine, metoprolol, and sulfamethoxazole, were detected in STP influent, STP effluent, and surface water. However, the mean of concentration was 561, 3305, 1805, 3900, 78, and 308 ng/L for atenolol, acetaminophen, theophylline, caffeine, metoprolol, and sulfamethoxazole, respectively, in STP influent.

The highest prescribed compounds detected in water samples were atenolol and metoprolol, with levels of 1009 and 606 ng/L, respectively. Non-prescription pharmaceuticals, caffeine, acetaminophen, and theophylline, were detected frequently and at high concentrations of 8700, 4919, and 2722 ng/L, respectively, in STP influent.

Supplementary Materials: The following are available online at http://www.mdpi.com/2073-4441/10/7/916/s1: Figure S1: Calibration curve graphs for all studied pharmaceuticals; Figure S2: LC chromatograms and mass spectra of some detected pharmaceuticals in STP influent.

Authors Contributions: F.F.A.-Q., Z.H.M. and N.A.T. initiated the research work. The method development, validation, and sample preparation were conducted by F.F.A.-Q. with contributions from Z.H.M., A.Y., N.A.T., and N.H. Funding Acquisition was provided by A.Y., N.A.T., N.H., and S.A. All authors read and approved the final manuscript.

Funding: This research received no external funding.

Acknowledgments: Support for this study was provided by Malaysia-Japan International Institute of Technology (MJIIT), Research Management Center (RMC), Universiti Teknologi Malaysia. The authors thank Mr. Alefee who is the person in-charge of LC-TOF/MS. The authors are thankful University of Babylon. Finally, the authors would like to thank all staff from ALIR lab, Universiti Kebangsaan Malaysia.

Conflicts of Interest: The authors declare no conflicts of interest.

References

1. De García, S.O.; García-Encina, P.A.; Irusta-Mata, R. The potential ecotoxicological impact of pharmaceutical and personal care products on humans and freshwater, based on USEtox™ characterization factors. A Spanish case study of toxicity impact scores. *Sci. Total Environ.* **2017**, *609*, 429–445. [CrossRef] [PubMed]
2. Langel, K.; Gunnar, T.; Ariniemi, K.; Rajamäki, O.; Lillsunde, P. A validated method for the detection and quantitation of 50 drugs of abuse and medicinal drugs in oral fluid by gas chromatography-mass spectrometry. *J. Chromatogr. B* **2011**, *879*, 859–870. [CrossRef] [PubMed]
3. Dobor, J.; Varga, M.; Yao, J.; Chen, H.; Palkó, G.; Záray, G. A new sample preparation method for determination of acidic drugs in sewage sludge applying microwave assisted solvent extraction followed by gas chromatography-mass spectrometry. *Microchem. J.* **2010**, *94*, 36–41. [CrossRef]
4. Li, X.S.; Li, S.; Wynveen, P.; Mork, K.; Kellermann, G. Development and validation of a specific and sensitive LC-MS/MS method for quantification of urinary catecholamines and application in biological variation studies. *Anal. Bioanal. Chem.* **2014**, *406*, 7287–7297. [CrossRef] [PubMed]
5. Bicker, J.; Fortuna, A.; Alves, G.; Falcão, A. Liquid chromatographic methods for the quantification of catecholamines and their metabolites in several biological samples—A review. *Anal. Chim. Acta* **2013**, *768*, 12–34. [CrossRef] [PubMed]
6. Jiang, L.; Chen, Y.; Chen, Y.; Ma, M.; Tan, Y.; Tang, H.; Chen, B. Determination of monoamine neurotransmitters in human urine by carrier-mediated liquid-phase microextraction based on solidification of stripping phase. *Talanta* **2015**, *144*, 356–362. [CrossRef] [PubMed]
7. Gu, Q.; Shi, X.; Yin, P.; Gao, P.; Lu, X.; Xu, G. Analysis of catecholamines and their metabolites in adrenal gland by liquid chromatography tandem mass spectrometry. *Anal. Chim. Acta* **2008**, *609*, 192–200. [CrossRef] [PubMed]
8. Mokh, S.; El Khatib, M.; Koubar, M.; Daher, Z.; Al Iskandarani, M. Innovative SPE-LC-MS/MS technique for the assessment of 63 pharmaceuticals and the detection of antibiotic-resistant-bacteria: A case study natural water sources in Lebanon. *Sci. Total Environ.* **2017**, *609*, 830–841. [CrossRef] [PubMed]
9. Paíga, P.; Lolić, A.; Hellebuyck, F.; Santos, L.H.; Correia, M.; Delerue-Matos, C. Development of a SPE-UHPLC-MS/MS methodology for the determination of non-steroidal anti-inflammatory and analgesic pharmaceuticals in seawater. *J. Pharm. Biomed.* **2015**, *106*, 61–70. [CrossRef] [PubMed]
10. Petrović, M.; Škrbić, B.; Živančev, J.; Ferrando-Climent, L.; Barcelo, D. Determination of 81 pharmaceutical drugs by high performance liquid chromatography coupled to mass spectrometry with hybrid triple quadrupole-linear ion trap in different types of water in Serbia. *Sci. Total Environ.* **2014**, *468*, 415–428. [CrossRef] [PubMed]
11. Al-Qaim, F.F.; Abdullah, M.P.; Othman, M.R.; Latip, J.; Afiq, W. A validation method development for simultaneous LC-ESI-TOF/MS analysis of some pharmaceuticals in Tangkas river-Malaysia. *J. Braz. Chem. Soc.* **2014**, *25*, 271–281. [CrossRef]
12. Malaysian Statistics on Medicine, Ministry of Health Malaysia, Kuala Lumpur. 2014. Available online: http://apps.who.int/medicinedocs/documents/s17580en/s17580en.pdf (accessed on 11 October 2017).
13. Drugbank Database. Available online: https://www.drugbank.ca/ (accessed on 28 February 2018).
14. World Health Organization (WHO). Available online: https://www.whocc.no/atc_ddd_index/?code=C07AB02&showdescription=yes (accessed on 28 February 2018).

15. Castiglioni, S.; Bagnati, R.; Calamari, D.; Fanelli, R.; Zuccato, E. A multiresidue analytical method using solid-phase extraction and high-pressure liquid chromatography tandem mass spectrometry to measure pharmaceuticals of different therapeutic classes in urban wastewaters. *J. Chromatogr. A* **2005**, *1092*, 206–215. [CrossRef] [PubMed]
16. Daneshvar, A.; Svanfelt, J.; Kronberg, L.; Prévost, M.; Weyhenmeyer, G.A. Seasonal variations in the occurrence and fate of basic and neutral pharmaceuticals in a Swedish river-lake system. *Chemosphere* **2010**, *80*, 301–309. [CrossRef] [PubMed]
17. Gros, M.; Petrović, M.; Barceló, D. Development of a multi-residue analytical methodology based on liquid chromatography-tandem mass spectrometry (LC-MS/MS) for screening and trace level determination of pharmaceuticals in surface and wastewaters. *Talanta* **2006**, *70*, 678–690. [CrossRef] [PubMed]
18. Yuan, S.; Jiang, X.; Xia, X.; Zhang, H.; Zheng, S. Detection, occurrence and fate of 22 psychiatric pharmaceuticals in psychiatric hospital and municipal wastewater treatment plants in Beijing, China. *Chemosphere* **2013**, *90*, 2520–2525. [CrossRef] [PubMed]
19. Al-Qaim, F.F.; Abdullah, M.P.; Othman, M.R.; Mussa, Z.H.; Zakaria, Z.; Latip, J.; Afiq, W.M. Investigation of the environmental transport of human pharmaceuticals to surface water: A case study of persistence of pharmaceuticals in effluent of sewage treatment plants and hospitals in Malaysia. *J. Braz. Chem. Soc.* **2015**, *26*, 1124–1135. [CrossRef]
20. Belay, A. Some biochemical compounds in coffee beans and methods developed for their analysis. *Int. J. Phys. Sci.* **2011**, *6*, 6373–6378.
21. Vieno, N.M.; Tuhkanen, T.; Kronberg, L. Analysis of neutral and basic pharmaceuticals in sewage treatment plants and in recipient rivers using solid phase extraction and liquid chromatography-tandem mass spectrometry detection. *J. Chromatogr. A* **2006**, *1134*, 101–111. [CrossRef] [PubMed]
22. Petrovic, M.; Gros, M.; Barcelo, D. Multi-residue analysis of pharmaceuticals in wastewater by ultra-performance liquid chromatography-quadrupole-time-of-flight mass spectrometry. *J. Chromatogr. A* **2006**, *1124*, 68–81. [CrossRef] [PubMed]
23. Ferrer, I.; Zweigenbaum, J.A.; Thurman, E.M. Analysis of 70 Environmental Protection Agency priority pharmaceuticals in water by EPA Method 1694. *J. Chromatogr. A* **2010**, *1217*, 5674–5686. [CrossRef] [PubMed]
24. Shaaban, H.; Górecki, T. Fast ultrahigh performance liquid chromatographic method for the simultaneous determination of 25 emerging contaminants in surface water and wastewater samples using superficially porous sub-3 μm particles as an alternative to fully porous sub-2 μm particles. *Talanta* **2012**, *100*, 80–89. [PubMed]
25. Hernando, M.D.; Petrovic, M.; Fernández-Alba, A.R.; Barceló, D. Analysis by liquid chromatography-electrospray ionization tandem mass spectrometry and acute toxicity evaluation for β-blockers and lipid-regulating agents in wastewater samples. *J. Chromatogr. A* **2004**, *1046*, 133–140. [PubMed]
26. Al-Qaim, F.F.; Yuzir, A.; Mussa, Z.H. Determination of theobromine and caffeine in some Malaysian beverages by liquid chromatography-time-offlight mass spectrometry. *Trop. J. Pharm. Res.* **2018**, *17*, 529–535. [CrossRef]
27. Pedrouzo, M.; Borrull, F.; Pocurull, E.; Marcé, R.M. Presence of pharmaceuticals and hormones in waters from sewage treatment plants. *Water Air Soil Pollut.* **2011**, *217*, 267–281. [CrossRef]
28. Huggett, D.B.; Khan, I.A.; Foran, C.M.; Schlenk, D. Determination of beta-adrenergic receptor blocking pharmaceuticals in United States wastewater effluent. *Environ. Pollut.* **2003**, *121*, 199–205. [CrossRef]
29. Kasprzyk-Hordern, B.; Dinsdale, R.M.; Guwy, A.J. Multiresidue methods for the analysis of pharmaceuticals, personal care products and illicit drugs in surface water and wastewater by solid-phase extraction and ultra performance liquid chromatography–electrospray tandem mass spectrometry. *Anal. Bioanal. Chem.* **2008**, *391*, 1293–1308. [CrossRef] [PubMed]
30. Ternes, T.A. Occurrence of drugs in German sewage treatment plants and rivers. *Water Res.* **1998**, *32*, 3245–3260. [CrossRef]

Article

Origin, Fate and Control of Pharmaceuticals in the Urban Water Cycle: A Case Study

Roberta Hofman-Caris [1], Thomas ter Laak [1], Hans Huiting [1], Harry Tolkamp [1,2], Ad de Man [3], Peter van Diepenbeek [4] and Jan Hofman [5,*]

1 KWR Watercycle Research Institute, PO Box 1072, 3430 BB Nieuwegein, The Netherlands; roberta.hofman-caris@kwrwater.nl (R.H.-C.); thomas.ter.laak@kwrwater.nl (T.t.L.); hans.huiting@kwrwater.nl (H.H.); harry.tolkamp@gmail.com (H.T.)

2 Waterschap Limburg, PO box 2207, 6040 CC Roermond, The Netherlands

3 Waterschapsbedrijf Limburg, PO Box 1315, 6040 KH Roermond, The Netherlands; addeman@wbl.nl

4 Waterleiding Maatschappij Limburg, PO Box 1060, 6201 BB Maastricht, The Netherlands; P.vanDiepenbeek@wml.nl

5 Department of Chemical Engineering, Water Innovation and Research Centre, University of Bath, Claverton Down, Bath BA2 7AY, UK

* Correspondence: j.a.h.hofman@bath.ac.uk; Tel.: +44-1225-383555

Received: 28 April 2019; Accepted: 15 May 2019; Published: 17 May 2019

Abstract: The aquatic environment and drinking water production are under increasing pressure from the presence of pharmaceuticals and their transformation products in surface waters. Demographic developments and climate change result in increasing environmental concentrations, deeming abatement measures necessary. Here, we report on an extensive case study around the river Meuse and its tributaries in the south of The Netherlands. For the first time, concentrations in the tributaries were measured and their apportionment to a drinking water intake downstream were calculated and measured. Large variations, depending on the river discharge were observed. At low discharge, total concentrations up to 40 μg/L were detected, with individual pharmaceuticals exceeding thresholds of toxicological concern and ecological water-quality standards. Several abatement options, like reorganization of wastewater treatment plants (WWTPs), and additional treatment of wastewater or drinking water were evaluated. Abatement at all WWTPs would result in a good chemical and ecological status in the rivers as required by the European Union (EU) Water Framework Directive. Considering long implementation periods and high investment costs, we recommend prioritizing additional treatment at the WWTPs with a high contribution to the environment. If drinking water quality is at risk, temporary treatment solutions in drinking water production can be considered. Pilot plant research proved that ultraviolet (UV) oxidation is a suitable solution for drinking water and wastewater treatment, the latter preferably in combination with effluent organic matter removal. In this way >95% of removal of pharmaceuticals and their transformation products can be achieved, both in drinking water and in wastewater. Application of UV/H_2O_2, preceded by humic acid removal by ion exchange, will cost about €0.23/m^3 treated water.

Keywords: pharmaceuticals; water quality; water treatment; wastewater treatment; abatement options

1. Introduction

Organic micropollutants in water have been a topic of interest for some time [1]. They include industrial compounds, pesticides, personal care products, steroid hormones and pharmaceuticals (both from human and veterinary consumption). The presence of pharmaceutical compounds in surface waters was suspected and proven already long ago [2,3]. More recently, it became apparent that many different pharmaceutical compounds are found in surface water and groundwater [4]. It is expected that the environmental numbers and concentrations will increase, because new pharmaceuticals are being developed and pharmaceutical consumption is increasing due to demographic changes such as growing and aging populations [5]. Another factor that probably will affect the pharmaceutical concentration in surface waters, is climate change, which in The Netherlands and Western Europe is expected to cause longer dry periods—the summer of 2018 was a good example—and thus higher concentrations of micropollutants in surface waters (less dilution) will occur [6,7].

Most of the pharmaceuticals and their metabolites are excreted from the body via urine and feces after use. They are transported to wastewater treatment plants (WWTPs), where they are discharged into the environment. To protect receiving surface waters, these WWTPs apply biological processes to remove organics, nitrogen and phosphorous from the wastewater, but they are not specifically designed to remove organic micropollutants, including pharmaceuticals. In practice, the removal of the organic micropollutants varies between 0% and 100%, with an average total removal of 60%–70%. As a result, significant pharmaceuticals concentrations are present in WWTP effluent and will end up in surface waters and, therefore, in sources for drinking water production [8].

As these biologically active compounds are designed to bring about a specific effect in organisms, there is an increasing understanding that the presence of pharmaceuticals in surface water is undesired [9–11]. It already has been shown that some of these compounds, like diclofenac, fluoxetine and hormone disruptors can change the behavior of predators in water and can accumulate through the food chain [12]. Others reported on fish feminization and reduced reproduction [9].

For drinking water, the risk for human health of individual compounds at low concentrations is negligible. However, little is known about the effects of long-term exposure to pharmaceutical mixtures. Furthermore, Dutch drinking water utilities have the policy of distributing impeccable water quality, which means that from a precautionary principle, pharmaceuticals and other micropollutants should in principle be absent in drinking water.

Because of the environmental concerns, the EU has developed a watch list that includes 17-beta-estradiol (E2), 17-alpha-ethinylestradiol (EE2), and diclofenac. It is expected that eventually standards will be set for these compounds [13,14]. The currently proposed revision of the European Union (EU) Drinking Water Directive [15], contains several new standards for chemical parameters, including beta-estradiol, and some endocrine disrupters.

Drinking water utilities and authorities for managing surface water quality are, therefore, looking for adequate solutions for protecting of surface water quality, sources for drinking water and the production of high-quality drinking water, free of pharmaceuticals and micropollutants. There are three approaches that can contribute to this: (1) prevent emissions by reducing consumption and removal at source, (2) removal during wastewater treatment, and (3) removal during drinking water treatment.

The first option would be the most elegant solution to prevent pharmaceuticals from entering the wastewater and the environment. However, people cannot be denied the use of medication. It would already have a large impact if the public and especially physicians would realize that pharmaceuticals are not harmless, and that in some cases it may be better to prescribe an alternative pharmaceutical or lower dose, to protect the environment [16,17]. This, however, will not result in the absence of pharmaceuticals and their metabolites in wastewater, and additional treatment in WWTP and/or drinking water treatment may become inevitable in future. Currently activated carbon adsorption or oxidation by ozone are considered as the current industry standard for this purpose, but these technologies come with high costs. Nanofiltration or reversed osmosis is not favored, as it is difficult to discharge the concentrates. Advanced oxidation processes (AOPs) do not have this disadvantage. The application of AOPs in wastewater treatment has been studied by several authors. In The Netherlands, where bromide concentrations on the average are about 120 μg/L in the Rhine and about 70 μg/L in the Meuse, where they enter the country [18], this would result in a significant increase in the bromate content of surface water. This is an unwanted side effect, as bromate is considered carcinogenic. Varanasi, Coscarelli et al. (2018) [19] studied ultraviolet (UV)/H_2O_2, UV/free chlorine and UV/persulfate processes (UV/PS) for the removal of trace organic compounds, and concluded that their efficiency is greatly affected by the presence of background dissolved organic matter (DOM). The performance of UV/H_2O_2 and UV/free chlorine processes is mainly affected by DOM containing more aliphatic components, whereas the performance of UV/PS mainly depends on the presence of aromatic compounds.

According to Nihemaiti et al. (2018) [20] UV/peroxydislufate (UV/PDS) processes are more effective than UV/H_2O_2 processes for the removal of trace organic compounds in pure water. However, electron-rich compounds in effluent organic matter (EfOM) will cause high competition in UV/PDS processes. Thus, the efficiency of UV/PDS strongly depended on the variation of the composition and concentration of DOM and nitrite. In general, higher UV fluences and oxidant doses were required to overcome the impact of the water matrix. Application of UV/H_2O_2 does not result in the formation of byproducts from H_2O_2, but in case of UV/PDS processes, the effect of the resulting sulfate concentration on salinity should be considered. Application of different AOPs for the treatment of municipal wastewater has also been evaluated by other authors [21–23]. In general UV/free chlorine processes show a better performance than UV/H_2O_2 processes, although UV/chlorine processes show a higher compound selectivity. The same conclusion was drawn by Guo et al. [24], who found that UV/chlorine processes are less affected by the water and wastewater matrices than UV/H_2O_2 processes (which are superior in pure water, if compounds are more sensitive towards hydroxyl radicals). However, in The Netherlands chlorine-based processes are not preferred because of the possible formation of chlorine containing byproducts. According to Miklos et al. [21], the efficiency of UV/H_2O_2 processes in wastewater treatment strongly depends on the nitrite concentration, and the matrix composition should be monitored for effective application of UV/H_2O_2 processes.

This paper presents a case study in the province of Limburg in The Netherlands, comprising the river Meuse and several tributaries, upstream from the drinking water production plant 'Heel', which uses the river water as a source. The study area comprises several wastewater treatment plants, with direct influence on the water quality of the intake of the drinking water plant. The origin and fate of pharmaceuticals and metabolites in the study area are assessed, as well as different abatement options for water-quality control. Data for this study were acquired from a broader national wastewater effluent survey and two pilot plant investigations. The novelty of the study is in its integrated character and bridging between drinking water production and source control. In The Netherlands, drinking water production and surface water management are traditionally separated and independent sectors.

The goal of this integrated study was to develop and compare practical and cost-effective solutions for tackling pharmaceuticals in the drinking water produced at Heel and the surface water quality in the study area and at the raw water intake. The criteria for evaluation are water quality, environmental benefits, required time for realization and costs.

2. Materials and Methods

2.1. General Approach and Data Used in This Study

The central point of the study was the drinking water production from surface water from the river Meuse at the treatment plant in Heel. The focus of this study is on the presence of pharmaceuticals and metabolites in the river water, and the direct influence of WWTP emissions in the upstream catchment area on the intake for the drinking water production. As abatement options, the application of advanced oxidation processes for drinking water or wastewater treatment were evaluated.

The evaluations and assessments done in this paper are based on the results of three underlying studies in collaboration with the drinking water company "Waterleiding Maatschappij Limburg", the water board "Waterschap Limburg", responsible for the surface water quality in the area, and the wastewater treatment operator "Waterschapsbedrijf Limburg". In addition, data from a national survey of effluent quality were used. Sampling and data collection took place in different time periods. Therefore, some differences in investigated contaminants occur in the different phases of this integrated study.

The approach followed in this study comprises of the following steps:

1. Assessment of concentrations of pharmaceuticals and metabolites in several wastewater treatment effluents throughout The Netherlands to provide contextual or reference data for comparison with the effluents in the study area.
2. Assessment of loads and concentrations of pharmaceuticals in the river Meuse and tributaries upstream of the drinking water intake of treatment plant Heel. This includes the apportionment of the contribution of pharmaceuticals of the different wastewater treatment plants in the study area to the drinking water intake.
3. Evaluating different abatement options for pharmaceuticals and metabolites in the study area, including options for drinking water treatment and wastewater treatment. The focus is on the application of advanced oxidation processes.
4. Finally, a vision for short term and longer-term solutions is presented.

The sections below describe the study area, the national sampling campaign, and the pilot plant studies for drinking water treatment and wastewater treatment with AOP.

2.2. Study Area and Sampling Points

The area studied in this paper is the river basin of the river Meuse, upstream from the water intake of the drinking water production plant Heel. The Meuse rises at the French Plateau de Langres and flows through Belgium, after which it enters The Netherlands at the town of Eijsden. The river basin upstream of the intake of water production plant Heel (WPH) counts about six million inhabitants, of which 5.3 million are in Belgium and France. For this research and drinking water production, the following tributaries are of importance: The Jeker (95% flowing through Belgium), the Geul (50% flowing through Belgium), the Geleenbeek (100% Netherlands), and the Slijbeek (partly Belgium).

An inventory was made of pharmaceuticals in surface water and in the effluent of several wastewater treatment plants (WWTPs) [25]. In winter 2011 samples were taken in 4 consecutive weeks at 6 different locations in the river Meuse and its tributaries in the southern part of Limburg and in a canal parallel to the Meuse (the "Lateraal Kanaal"). Sample locations (See Figure 1b) were ① Eijsden (near the Belgian-Dutch border), ② the river Jeker in Maastricht, near the place where it enters the Meuse, ③ the river Geul near Meerssen, ④ the Geleenbeek near Oud Roosteren, ⑤ the Slijbeek, near the place where it enters the Lateraal Kanaal, and ⑥ at the intake of the drinking water production site Heel, which uses surface water from the Lateraal Kanaal as a source. The samples were single grab samples taken from bridges. The first two samples were taken at the end of a very long dry period (6 months) with a very low discharge flow in the Meuse. After the second sampling, rainfall commenced, the discharge flow increased. The third and fourth samples were, therefore, taken during a higher river discharge. See Supplementary Information for details Figures S1–S7 for details of the sampling points and the hydrograph.

2.3. National Sampling Campaign

In the national sampling campaign, effluent samples were taken at the WWTPs of Garmerwolde (A), Utrecht (B), Rotterdam (C), Eindhoven (D), Panheel (E) and Roermond (F) (see Figure 1a). Samples were single samples collected with an automatic sample collector over 24 h and proportional to the effluent flow rate, under dry weather conditions. In the WWTP effluent total organic carbon (TOC), dissolved organic carbon (DOC), and UV-Transmittance (UV-T254) at 254 nm were measured at the KWR laboratory. Composition of effluent organic matter (EfOM) was analyzed by DOC-Labor Dr. Huber (Eisenbahnstr. 6, 76229 Karlsruhe, Germany), applying an LC-OCD method [26]. The pharmaceuticals studied were selected based on consumption, occurrence in the environment, physico-chemical properties, the availability of standards and analytical methods [27]. Concentrations of pharmaceuticals and metabolites in surface water and WWTP effluents were measured according to the ultra-high performance liquid chromatography–tandem mass spectrometry (UPLC-MS/MS) method described previously Wols et al. [28].

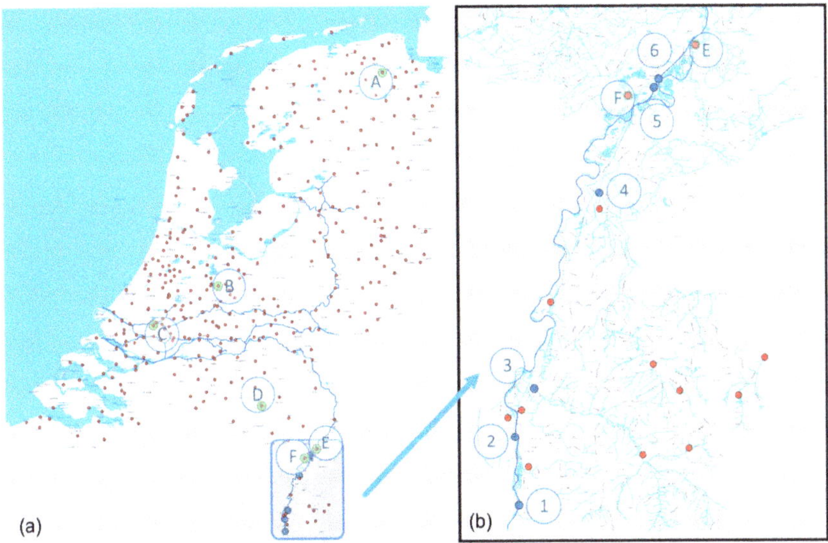

Figure 1. (**a**): Wastewater treatment plants (WWTPs) in The Netherlands (red dots) and WWTPs sampled for effluent characterization (green-red dots, indicated A to F); (**b**): Sample locations in surface water bodies (blue dots, indicated 1 to 6) in the southern part of the province of Limburg.

2.4. Wastewater Treatment Pilot Plant Panheel

To study the effect of additional treatment, experiments were carried out in a pilot set-up at WWTP Panheel, in which a mixture of more than 40 pharmaceuticals was dosed to the effluent to be able to study the removal efficiency of the additional treatment. The compounds were selected based on their presence in WWTP effluents and surface waters. The pilot plant comprised of a multi-layer filter (proprietary filter material; PureBlue Water, Kapellebrug, The Netherlands) to remove suspended solids and particles, an anion exchange (IEX) unit containing Lewatit S6368A resin (Lanxess, Brussels, Belgium) to remove humic acids, and a UV reactor (Type M3 by PureBlue Water; Kapellebrug, The Netherlands), which, after addition of H_2O_2 to the solution, was operated at a dose of 150 or 300 mJ/cm^2 for a UV/H_2O_2 process. The UV reactor was equipped with a low-pressure Amalgam lamp of 90 W. A largely over-dimensioned granular activated carbon (GAC) filtration unit filled with Norit GAC (type PK 1-3; Cabott Norit Netherlands, Amersfoort, The Netherlands) was used to prevent any discharge of added pharmaceuticals to the surface water. Further details on the pilot set-up can be found in Hofman-Caris et al. [29].

2.5. Drinking Water Treatment Pilot Plant Heel

The experiments at drinking water treatment plant Heel were carried out in a pilot set-up as described by Hofman-Caris et al. [27]. The flow through the reactor was 1–2.5 m^3/h. The UV reactor (type D200, Van Remmen UV-Techniek, Wijhe, The Netherlands) was optimized for advanced oxidation reactions according to [30]. It had been equipped with one LP UV lamp (Heraeus NNI 125-84-XL, Hanau, Germany) and two baffles to improve the flow conditions. Here too the mixture of over 40 pharmaceuticals and a H_2O_2 solution were added to the influent of the UV reactor. Again, a largely over-dimensioned granular activated carbon (GAC) filtration unit was used to prevent any discharge of added pharmaceuticals to the environment.

3. Results

3.1. Analysis of Wastewater Treatment Plant (WWTP) Effluents

The presence of pharmaceuticals in various WWTP effluents throughout The Netherlands is shown in Figure 2.

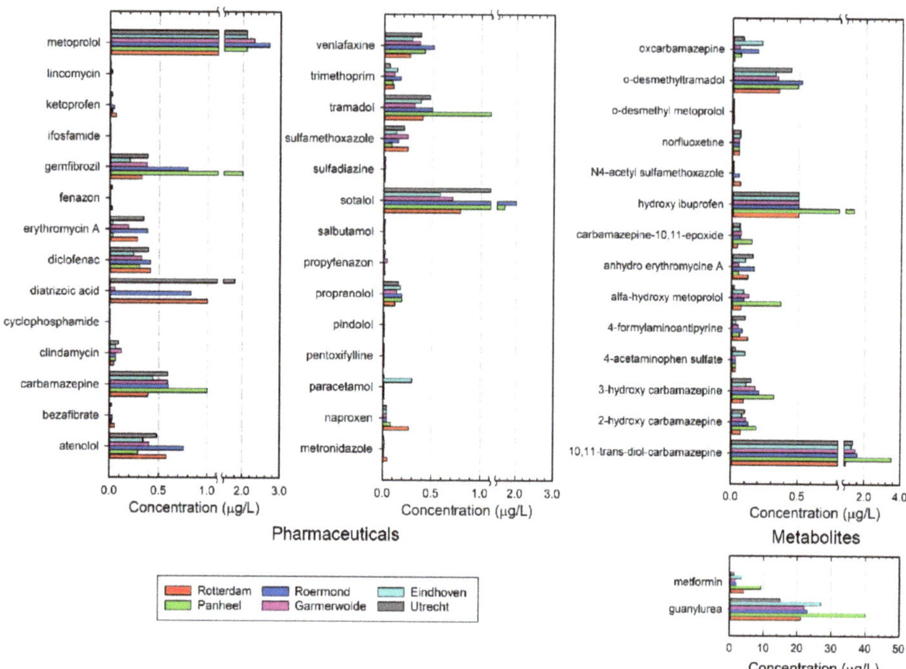

Figure 2. Concentrations of pharmaceuticals (left two graphs) and metabolites (right graph) in WWTP effluents in The Netherlands. The concentrations of metformin and its metabolite guanylurea are present in significantly higher concentrations, and therefore shown separately in the lower right graph.

From these data it can be concluded that there are large differences in the presence of certain pharmaceuticals in the effluent of various WWTPs. These differences can have several causes: e.g., sampling conditions, and diurnal and seasonal patterns can affect the observed concentrations. Removal efficiencies may vary too. Some compounds, like sotalol, were found in relatively high concentrations in the effluent of Utrecht, Roermond and Panheel (1.5–2.0 μg/L), whereas concentrations were smaller in Eindhoven, Garmerwolde and Rotterdam (0.5–0.8 μg/L). In contrast, diclofenac (0.24–0.41 μg/L) and venlafaxine (0.27–0.51 μg/L) had similar concentrations in all WWTP effluents analyzed. Tramadol, gemfibrozil, and carbamazepine were found in relatively high concentrations in the Panheel effluent, whereas in Utrecht a high concentration of diatrizoic acid was observed. In general, the effluents contained 14–28 μg/L of total pharmaceuticals and metabolites included in the analytical procedure, the average total pharmaceuticals concentration being about 16 μg/L. In the Panheel effluent the average concentration, however, was about 28 μg/L, which is relatively high. This is probably caused by the relatively high contribution of a nursing home, which sends its wastewater to this rather small WWTP.

Such factors may also affect the presence of individual pharmaceuticals in the effluent, which confirms earlier findings of the ZORG project in The Netherlands [31].

The data also indicate that some pharmaceuticals can easily be removed by the WWTP (like paracetamol, sulfadiazine, cyclophosphamide, norfluoxetine, lincomycin, phenazone, cyclophosphamide, bezafibrate, sulphadiazine, salbutamol, propyphenazone, pentoxyfylline and metronidazole), whereas others are very difficult to remove (like diatrizoic acid, metoprolol and diclofenac). Carbamazepine is partly converted, either by the human metabolism or by biodegradation. Several transformation products of carbamazepine (oxcarbamazepine, carbamazepine-10,11-epoxide, 2-hydroxy carbamazepine, 3-hydroxy carbamazepine, and 10,11-trans-diol carbamazepine) could be detected. Also, for tramadol and ibuprofen, transformation products can be found. Metformin is an antidiabetic that is present in high concentrations in wastewater because it is prescribed in high daily dosages to a large part of the population. In the WWTP metformin is converted by means of biodegradation into guanylurea, which is very difficult to further degrade in a WWTP. As a result, the concentrations of metformin and guanylurea are very high: in general, their concentrations are above 10 μg/L; in Panheel concentrations up to 40 μg/L were observed. Although often considered relatively harmless, metformin can act as an endocrine disruptor at environmentally relevant concentrations [32,33].

3.2. Pharmaceutical Loads in Surface Waters

The concentrations found in the WWTPs are a strong indication that WWTP effluent significantly contributes to the pharmaceutical load in surface waters. This is confirmed by the pharmaceuticals' concentrations found in the Meuse and its tributaries, as shown in Figure 3. Here the average concentration of four measurements (two in November, and two in December) is shown. The full data set can be found in the Supplementary Information (Tables S1–S12).

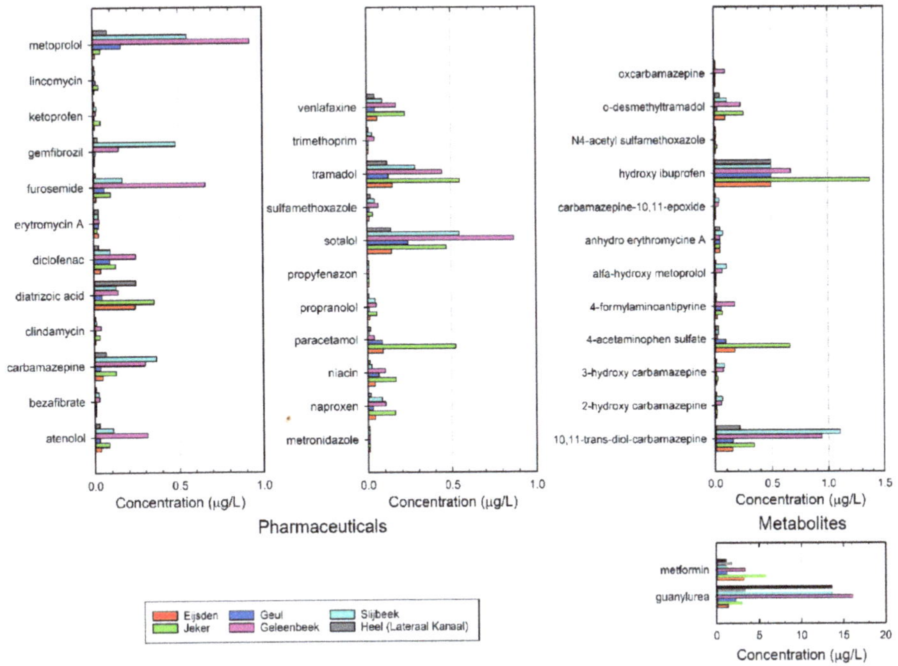

Figure 3. Average concentrations (n = 4) of pharmaceuticals and their metabolites detected in the River Meuse and its tributaries.

The pharmaceuticals load in the Meuse crossing the border between Belgium and The Netherlands strongly depends on the river discharge, which can vary almost two orders of magnitude. The minimum discharge at Eijsden is about 10 m^3/s, the nominal flow 147 m^3/s, whereas the maximum discharge is about 1400 m^3/s (period 2005–2009). In the most extreme situations, the discharge can even be as high as 3000 m^3/s. As the discharge of the Meuse depends on the rainfall in its catchment area, dry periods with low discharge up to six months may occur. The pharmaceutical load of the Meuse in Eijsden was calculated from the measured concentrations and the river discharge of 14 m^3/s (November 2011) and of 253 m^3/s (December 2011) (See discharge graph in the supplementary information, Figure S7). This resulted in an average load of 10 kg/day (standard deviation 18%) under low discharge conditions, and of 106 kg/day (standard deviation 32%) under higher discharge conditions. The load in the high-flow condition is higher than one would expect based on dilution. This is most probably caused by a 'first-flush' effect. The concentrations were measured at the initial flank of a rapidly increasing discharge: the flow was already high, but water with high concentrations was pushed forward in the river, without dilution.

Figure 3 shows the pharmaceuticals concentrations present in the various tributaries. In the Jeker, Geleenbeek and Slijbeek concentrations were higher than in the Meuse, due to the high load of WWTP effluent in these tributaries. WWTP effluent forms a significant part of the total discharge in these tributaries, as shown in Table 1. The WWTP of Panheel (25,000 PE) discharges to the Slijbeek and the WWTPs of Susteren (216,500 PE), Hoensbroek (240,000 PE) and Heerlen (65,600 PE) discharge to the Geleenbeek, resulting in relatively high pharmaceutical concentrations. In the Geleenbeek and the Slijbeek, the river discharge originated for roughly 40–50% from effluent. The amount of wastewater in the Jeker could not be quantified because of an unknown but significant contribution of untreated municipal wastewater from Belgium [34]. It is estimated that the wastewater of about 120,000 Belgian inhabitants is discharged at the Jeker. About one third of this had not or limitedly been treated (Situation 2011). Instead, a bacterial suspension is added to the river to improve water quality. Pharmaceutical concentrations in the Jeker appear to be a little lower than in the Geleenbeek and the Slijbeek.

Metformin is converted to guanylurea by biodegradation during biological wastewater treatment. The fact that in the Jeker at the time of the measurement a relatively large contribution of untreated wastewater was found probably accounts for the high contribution of metformin compared with its metabolite guanylurea. Furthermore, the concentrations of some painkillers and their transformation products (tramadol, paracetamol, acetaminophen sulphate, and ibuprofen) are higher in the Jeker than in other surface waters. This may be related to differences in the pharmaceutical use between The Netherlands and Belgium but can also result from a larger contribution of untreated wastewater, in which these compounds were not removed. The concentration in the Geul is the lowest, because of the relatively low contribution of effluent on the river discharge.

Ter Laak et al. [35] combined data on the flow of different rivers and demographics of the catchments to calculate daily per capita loads of pharmaceuticals and metabolites. Subsequently, they linked these loads to sales data of pharmaceuticals in the catchment and, considering human excretion and removal by WWTPs, thus were able to predict actual loads within a factor of three for most pharmaceuticals. Furthermore, from this study it became clear that there are differences in the use of pharmaceuticals per capita in The Netherlands and Belgium: the Jeker contains relatively high concentrations of metformin, diatrizoic acid, venlafaxine, tramadol, paracetamol, niacin, and naproxen. Also, the study by Ter Laak et al. [35] revealed differences in prescription practice between the two countries. Furthermore, the river is strongly affected by WWTP effluent and—at the time of the research—untreated municipal wastewater from Belgium.

Table 1. Most common pharmaceuticals and metabolites detected in the River Meuse and its tributaries in The Netherlands, expressed as the average contribution (%) to the total load of pharmaceuticals and transformation products analyzed.

	Meuse 'Low'	**Meuse 'High'**	**Geul**	**Geleenbeek**	**Slijbeek**	**Jeker**
Discharge (m^3/s)	14 [1] Near minimum	253 [2] Above nominal	2.8 [3] Nominal	2.1 [3] Nominal	0.1 [4] Estimated	1.7 [3] Nominal
Proportion of WWTP effluent to discharge (%)	No data available	No data available	15	38	51	Unknow; untreated discharges
Total pharmaceuticals load (kg/day)	10	106	1.4	4.9	0.1	2.2
guanylurea	25%	12%	40%	60%	68%	20%
metformine	47%	49%	22%	13%	6%	39%
10,11-trans-diol-carbamazepine	2%	2%	3%	4%	5%	2%
metoprolol	0%	0%	3%	3%	3%	0%
sotalol	2%	2%	4%	3%	3%	3%
hydroxy ibuprofen	6%	10%	9%	3%	2%	9%
furosemide	0%	0%	1%	2%	1%	1%
tramadol	2%	2%	2%	2%	1%	4%
atenolol	1%	1%	0%	1%	1%	1%
carbamazepine	1%	1%	1%	1%	2%	1%
diclofenac	1%	1%	2%	1%	0%	1%
Other 26 compounds	13%	19%	13%	7%	8%	19%

[1] November 2011; [2] December 2011; [3] Measurements every 15 min, 1-1-2008 to 21-12-2011; [4] Estimated by the Water Authority 'Peel en Maasvallei'.

3.3. *Abatement Options to Protect Surface Water Quality and Drinking Water Production*

The estimated contribution of the Meuse and its tributaries to the concentration of pharmaceuticals and metabolites in the intake of drinking water production site Heel is shown in Figure 4. The darker shaded bars for November and December are based on loads predicted from the actual measurement in these months. The lighter bars are based on load predictions for extreme low, median and extreme high discharge. As expected, the river Meuse has the largest contribution to the intake, especially in median- and high-discharge situations. At extreme high flow it is also observed that the concentrations will be low due to dilution. However, during low flow, which can take up to 6 months in dry years, the direct contributions of the tributaries to the drinking water intake are large, up to almost 50%. Treatment measures on the WWTPs along the tributaries can effectively reduce the pharmaceutical compounds in the intake for drinking water production.

As it is expected that the pharmaceutical loads and concentrations will increase in the future, several abatement options were studied to control surface water quality and/or prevent concentrations that are too high in drinking water.

Table 2 gives an overview of the possibilities, their expected effects and the estimated operational costs. The options will be described in more detail below. The first abatement option in Table 2 is reducing the emission of pharmaceuticals at the source. This is not easy to realize. People need pharmaceuticals because of health problems, and it is not ethical to abstain pharmaceuticals only because of environmental reasons. In many cases, alternative pharmaceuticals will probably show comparable behavior in the environment, as the activity of a compound is strongly related to its chemical structure, which, as a result, will show strong resemblances. In some cases, physicians can prescribe different pharmaceuticals (like naproxen instead of diclofenac) or minimize doses.

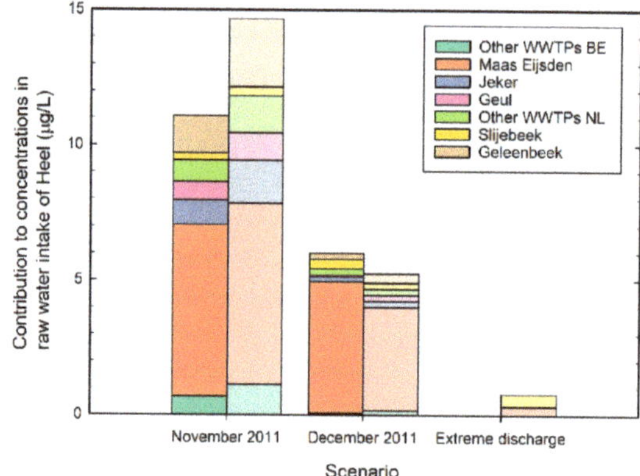

Figure 4. Contribution of several sources to the pharmaceutical and metabolite concentration at the intake of drinking water production site Heel. The dark shaded bars are based on load predictions from the actual measurements of concentrations and discharges in November and December 2011. The lighter shaded bars are generalized load predictions for low (comparable to November), median (comparable to December) and extremely high discharge.

Table 2. Potential abatement options to improve quality of surface waters and the source water for drinking water treatment at the production site Heel.

Abatement Option	Effect on Drinking Water	Effect on Surface Water	Total Additional Costs [1]	Advantages	Disadvantages
Prevent and reduce pharmaceutical emission at source (toilets)	Effective	Effective	unknown	Protects aquatic environment	Difficult to realize; long-term effects
Different layout of water system; diverting WWTP effluent downstream of the drinking water intake	Effective	Not effective	low	Quick solution	Emergency measure
Extension of individual WWTPs	Not effective unless realized on multiple locations	Effective for small surface waters	8–15 M€/year	Local improvement of surface water quality	Only effective for drinking water on long-term and at large-scale application
Extension of WWTPs on a large, international scale	Effective	Effective	8–15 M€/year in The Netherlands	Strong improvement of surface water quality	Long-term realization
Extension drinking water treatment	Effective	Not effective	4–8 M€/year	Short-term realization	No improvement of surface water quality

[1] Costs are calculated based on the cost standard and related calculator, developed by the Dutch water sector. See www.kostenstandaard.nl.

The second option is to divert effluent by realizing a different layout of the water system, combining some WWTPs, or relocating them downstream of the drinking water intake. This may be effective for the quality improvement of the local intake water of the drinking water production, but for the total aquatic environment, it is not effective, as the same load of pharmaceuticals and metabolites eventually will end up in the surface water. Moreover, there will be no change for drinking water intakes further downstream. For the situation in Heel this could be an option that is easy to realize, because WWTP Panheel discharges via the Slijbeek into the Lateraal Kanaal, a few hundred meters upstream of the raw water intake of WPH.

The third option, extension of individual WWTPs with additional treatment steps, will be very effective for the local aquatic environment [36]. Some pilot experiments were carried out at WWTP Panheel [29]. It was shown that the effectiveness of advanced oxidation, e.g., based on UV/H_2O_2

processes, can very much be improved by first removing the humic acid part of the effluent organic matter by means of ion exchange. In general, UV/H_2O_2 processes have a high energy demand, but removal of the humic acids increased the UV-transmission from 38% to 85% resulting in an 84% reduced energy demand. Furthermore, most pharmaceuticals were broken down >85% (Figure 5). Interestingly, a UV dose of 150 mJ/cm^2, combined with a H_2O_2 concentration of 10 mg/L, seems to be quite effective. This is a much lower dose than commonly applied for advanced oxidation of contaminants in drinking water (often about 500 mJ/cm^2). Hence, the costs of UV/H_2O_2, preceded by humic acid removal by ion exchange, will be relatively low at about €0.23/m^3 treated water.

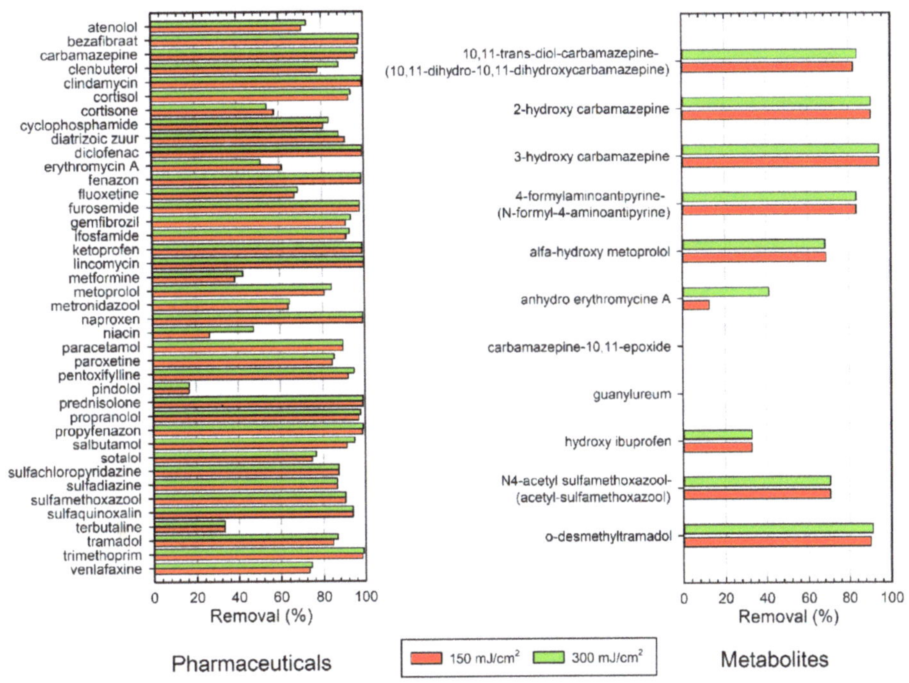

Figure 5. Removal of pharmaceuticals (left graph) and metabolites (right graph) in the effluent of WWTP Panheel, by applying ultraviolet (UV)/H_2O_2 at a UV dose of 150 and 300 mJ/cm^2.

The main disadvantage of this system is the fact that it only results in a very local improvement of the aquatic environment, and thus only will be effective if it is applied at a large scale, involving many or all WWTPs. Furthermore, in the actual situation in the study area, the largest load of pharmaceuticals is present in the Meuse already before it enters The Netherlands. Therefore, only adjusting the Dutch WWTPs will not be very effective at improving the total surface water quality. Nevertheless, the contribution of the tributaries can be significant in dry periods as can be seen in Figure 4. These dry periods are more likely to occur in future and last longer, due to climate change, and these dry periods result in the highest concentrations and associated risk [7]. To obtain an effective reduction of the load all, or at least the majority, of WWTPs in the catchment area require additional treatment. Depending on what needs protection, smart abatement solutions can be developed [37]. This is indicated by the fourth option in Table 2. In that case the total surface water quality, and thus also the source water quality for drinking water production would be significantly improved. However, this will involve considerable investments and solutions have to be found in international cooperation, since the rivers cross borders.

The final solution, option five, may be additional treatment at the point of drinking water production. This has been tested for the situation in Heel, applying advanced oxidation (UV/H_2O_2) [27]. As the UV-transmission of the intake water already is high (94%) due to the river bank filtration, this process is very effective, removing >90% of most pharmaceuticals (Figure 6). By optimizing the reactor geometry, it was shown that even in this case significant improvement of the process efficiency (a 40% decrease in energy demand) could be obtained. Due the very high UV transmittance of the influent of the UV reactor, reflection of UV irradiation at the outer reactor wall occurred. As a result, in the pilot set-up applied, the actual UV dose could not be decreased below 365 mJ/cm^2, but obviously, for practical applications a lower UV dose would have been sufficient, as most pharmaceuticals were removed to a high degree. Metabolites in general were also removed, although for 3-hydroxy carbamazepine formation of about 25 µg/L could be observed, probably due to the conversion of carbamazepine by the Advanced Oxidation Processes).

A drawback of this fifth option is that, although it can significantly improve drinking water quality in the case of increasing pharmaceutical concentrations, it does not affect surface water quality and thus the aquatic environment.

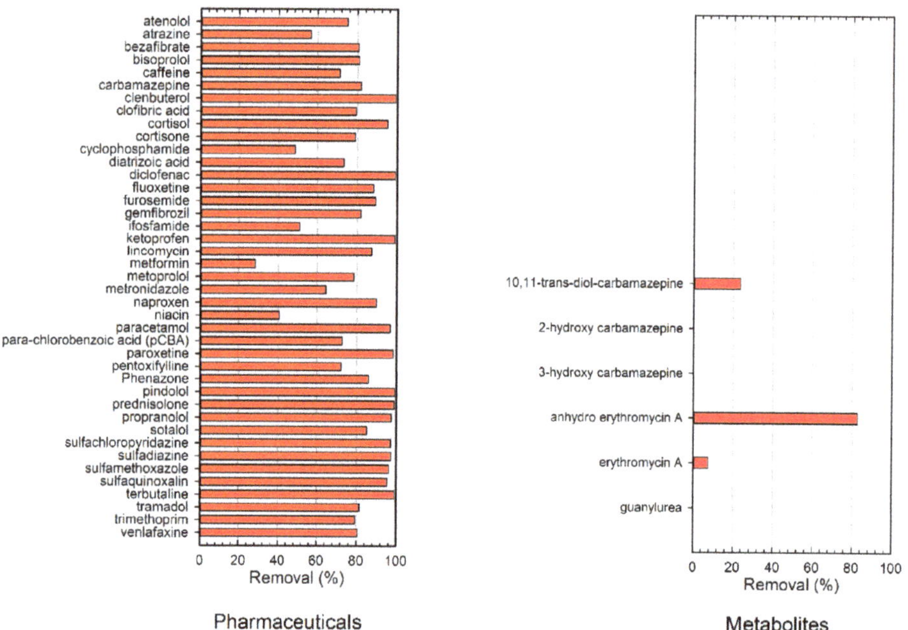

Figure 6. Removal of pharmaceuticals (left graph) and metabolites (right graph) at drinking water production site Heel, applying a UV dose of 365 mJ/cm^2 and a H_2O_2 concentration of 10 mg/L.

4. Discussion

The above case study in Limburg has indicated that pharmaceuticals and metabolites can be present in surface water in relatively high concentrations in the study area. The concentrations are largely influenced by the river discharge flow and are expected to increase in future due to demographic changes and prolonged dry periods. In these dry periods in particular, concentrations are high in the tributaries and the raw water intake of the drinking water production. This is an undesired situation from a precautionary perspective for drinking water quality and aquatic life.

Different abatement options have been considered in this case study ranging from changes in the water system to the use of additional treatment technology for drinking water and wastewater treatment. It was concluded that the best measure would be to remove pharmaceuticals on all WWTPs in the catchment. This would reduce concentrations of pharmaceuticals in the river system significantly, creating a good chemical and ecological status, and it would make additional treatment in the drinking water production superfluous. In this way the system could comply perfectly with the regulations of the Water Framework Directive. However, the investment costs would be high and international collaboration is required, probably making this a long-term solution [36,38].

The methodology presented here to identify the origin and fate of pharmaceutical compounds can be used to identify WWTPs that act as hotspots for pollution. At these hot-spots, additional treatment technology could be installed with priority to improve the surface water quality in the receiving water bodies. If during the development of this abatement option the drinking water quality still would be at risk, temporary additional treatment at the drinking water production point could be considered. These temporary measures should always be considered in an integrated approach and should not be used as an excuse for delaying treatment at WWTP hotspots.

The UV/H_2O_2 technology tested at the two sites in this case study is extremely suitable, either in combination with anion exchange for EfOM removal as a permanent solution at WWTPs, or as a temporary solution for drinking water if the water quality is at risk. The technology has a small footprint, can be built in a modular fashion, and has a high removal efficiency for pharmaceutical compounds. Furthermore, it can be operated at reasonable cost.

5. Conclusions

This case study has indicated and confirmed that pharmaceuticals and their metabolites are present throughout the study area, originating from WWTP discharge. It has been shown that significant contributions from effluents can be observed in smaller tributaries. The WWTP effluent contributes to a large extent to the discharge and pollutant load. In our study it was observed that the top 10 of the highest concentrations of pharmaceuticals and metabolites determine 90% of the total load.

To tackle the challenge of pharmaceuticals in the water cycle, we envision a combined approach with a long-term approach to remove pharmaceutical compounds form WWTP effluent, with priority on hotspots. If the drinking water quality is at risk, a temporary treatment solution for the drinking water production with a small-footprint and modular flexible design can be considered. Based on our pilot plant results, both systems the use of UV oxidation could be a viable solution. On the WWTPs, this can be combined with EfOM removal to improve efficiency.

Supplementary Materials: The following are available online at http://www.mdpi.com/2073-4441/11/5/1034/s1, Figure S1: Sampling location Meuse at Eijsden, Figure S2: Sampling point Jeker at Maastricht, Figure S3: Sampling point Geul at Meerssen, Figure S4: Sampling point Geleenbeek at Oud-Roosteren, Figure S5: Sampling point Slijbeek, Figure S6: Sampling point at intake of Water Treatment Plant Heel, Lateraalkanaal, Figure S7: Discharge of the Meuse during 2011. The river showed an extreme low flow from May to early December. The black dots indicate the sampling dates for pharmaceuticals, Table S1: Concentrations of pharmaceuticals in the Lateraal Kanaal, measured at four different dates in 2011, Table S2: Concentrations of pharmaceuticals in the Slijbeek, measured at four different dates in 2011, Table S3: Concentrations of pharmaceuticals in the Geleenbeek, measured at four different dates in 2011, Table S4: Concentrations of pharmaceuticals in the Geul, measured at four different dates in 2011, Table S5: Concentrations of pharmaceuticals in the Jeker, measured at four different dates in 2011, Table S6: Concentrations of pharmaceuticals in the Meuse at Eijsden, measured at four different dates in 2011, Table S7: Concentrations of metabolites in the Lateraal Kanaal, measured at four different dates in 2011, Table S8: Concentrations of metabolites in the Slijbeek, measured at four different dates in 2011, Table S9: Concentrations of metabolites in the Geleenbeek, measured at four different dates in 2011, Table S10: Concentrations of metabolites in the Geul, measured at four different dates in 2011, Table S11: Concentrations of metabolites in the Jeker, measured at four different dates in 2011, Table S12: Concentrations of metabolites in the Meuse at Eijsden, measured at four different dates in 2011.

Author Contributions: Conceptualization, R.H.-C., T.t.L. and J.H.; Formal analysis, R.H.-C., T.t.L. and J.H.; Investigation, R.H.-C., T.t.L., H.H., H.T., A.d.M., P.v.D. and J.H.; Methodology, R.H.-C., T.t.L., H.H., H.T., A.d.M., P.v.D. and J.H.; Validation, R.H.-C.; Visualization, J.H.; Writing – original draft, R.H.-C.; Writing – review and editing, R.H.-C., T.t.L., H.H., H.T., P.v.D. and J.H.

Funding: This project has been co-financed with PPS-funding from the Topconsortia for Knowledge and Innovation (TKI's) of the Ministry of Economic Affairs and Climate.

Acknowledgments: The authors would like to thank Waterleiding Maatschappij Limburg and Waterschapsbedrijf Limburg for providing data and their support for conducting the pilot plant research.

Conflicts of Interest: The authors declare no conflict of interest.

References

1. Barbosa, M.O.; Moreira, N.F.F.; Ribeiro, A.R.; Pereira, M.F.R.; Silva, A.M.T. Occurrence and removal of organic micropollutants: An overview of the watch list of EU Decision 2015/495. *Water Res.* **2016**, *94*, 257–279. [CrossRef]
2. Richardson, M.L.; Bowron, J.M. The fate of pharmaceutical chemicals in the aquatic environment. *J. Pharm. Pharmacol.* **1985**, *37*, 1–12. [CrossRef]
3. Daughton, C.G.; Ternes, T.A. Pharmaceuticals and personal care products in the environment: Agents of subtle change? *Environ. Health Perspect.* **1999**, *107*, 907–938. [CrossRef] [PubMed]
4. Sui, Q.; Cao, X.; Lu, S.; Zhao, W.; Qiu, Z.; Yu, G. Occurrence, sources and fate of pharmaceuticals and personal care products in the groundwater: A review. *Emerg. Contam.* **2015**, *1*, 14–24. [CrossRef]
5. Van Der Aa, N.G.F.M.; Kommer, G.J.; Van Montfoort, J.E.; Versteegh, J.F.M. Demographic projections of future pharmaceutical consumption in the Netherlands. *Water Sci. Technol.* **2011**, *63*, 825–831. [CrossRef]
6. Klein Tank, A.; Beersma, J.; Bessembinder, J.; van den Hurk, B.; Lenderink, G. *KNMI '14; Klimaatscenario's Voor Nederland, Leidraad Voor Professionals in Klimaatadaptatie*; 2015 Revised Version; Royal Netherlands Meteorological Institute KNMI: De Bilt, The Netherlands, 2015. (In Dutch)
7. Sjerps, R.M.A.; Ter Laak, T.L.; Zwolsman, G.J.J.G. Projected impact of climate change and chemical emissions on water quality of the European rivers Rhine and Meuse: A drinking water perspective. *Sci. Total Environ.* **2017**, *601*, 1682–1694. [CrossRef]
8. Heberer, T. Occurrence, fate, and removal of pharmaceutical residues in the aquatic environment: A review of recent research data. *Toxicol. Lett.* **2002**, *131*, 5–17. [CrossRef]
9. Kidd, K.A.; Blanchfield, P.J.; Mills, K.H.; Palace, V.P.; Evans, R.E.; Lazorchak, J.M.; Flick, R.W. Collapse of a fish population after exposure to a synthetic estrogen. *Proc. Natl. Acad. Sci. USA* **2007**, *104*, 8897–8901. [CrossRef]
10. Crane, M.; Watts, C.; Boucard, T. Chronic aquatic environmental risks from exposure to human pharmaceuticals. *Sci. Total Environ.* **2006**, *367*, 23–41. [CrossRef]
11. Santos, L.H.; Araújo, A.N.; Fachini, A.; Pena, A.; Delerue-Matos, C.; Montenegro, M.C.B.S.M. Ecotoxicological aspects related to the presence of pharmaceuticals in the aquatic environment. *J. Hazard. Mater.* **2010**, *175*, 45–95. [CrossRef]
12. Brodin, T.; Piovano, S.; Fick, J.; Klaminder, J.; Heynen, M.; Jonsson, M. Ecological effects of pharmaceuticals in aquatic systems—Impacts through behavioural alterations. *Philos. Trans. R. Soc. B Biol. Sci.* **2014**, *369*. [CrossRef] [PubMed]
13. European Commission. Commission Implementing Decision (EU) 2015/495 of 20 March 2015 Establishing a Watch List of Substances for Union-Wide Monitoring in the Field of Water Policy Pursuant to Directive 2008/105/EC of the European Parliament and of the Council. Available online: http://eur-lex.europa.eu/legal-content/EN/TXT/PDF/?uri=CELEX:32015D0495&from=EN (accessed on 22 October 2018).
14. Carvalho, R.N.; Ceriani, L.; Ippolito, A.; Lettieri, T. *Development of the first Watch List under the Environmental Quality Standards Directive*; EC Joint Research Centre: Ispra, Italy, 2015.
15. European Commission. Proposal for a Directive of the European Parliament and of the Council on the Quality of Water Intended for Human Consumption (Recast). Available online: http://eur-lex.europa.eu/resource.html?uri=cellar:8c5065b2-074f-11e8-b8f5-01aa75ed71a1.0016.02/DOC_1&format=PDF (accessed on 22 October 2018).
16. Van der Grinten, E.; van der Maaden, T.; van Vlaardingen, P.L.A.; Venhuis, B.J.; Moermond, C.T.A. *Milieuafwegingen in de Geneesmiddelvoorziening*; Rijksinstituut voor Volksgezondheid en Milieu; Ministerie van Volksgezondheid, Welzijn en Sport: Bilthoven, The Netherlands, 2017. (In Dutch)
17. Waterschap Drents Overijsselse Delta. *Evaluatie van het Project Milieubewust Medicijnen Voorschrijven Door Huisartsen in Meppel*; Waterschap Drents Overijsselse Delta: Meppel, The Netherlands, 2017. (In Dutch)

18. Mulder, M.; Antakyali, D.; Ante, S. *Verwijdering van Microverontreinigingen uit Effluenten van RWZI's: Een Vertaling van Kennis en Ervaring uit Duitsland en Zwitserland.*; Stichting Toegepast Onderzoek Waterbeheer (STOWA): Amersfoort, The Netherlands, 2015. (In Dutch)
19. Varanasi, L.; Coscarelli, E.; Khaksari, M.; Mazzoleni, L.R.; Minakata, D. Transformations of dissolved organic matter induced by UV photolysis, Hydroxyl radicals, chlorine radicals, and sulfate radicals in aqueous-phase UV-Based advanced oxidation processes. *Water Res.* **2018**, *135*, 22–30. [CrossRef]
20. Nihemaiti, M.; Miklos, D.B.; Hübner, U.; Linden, K.G.; Drewes, J.E.; Croué, J.P. Removal of trace organic chemicals in wastewater effluent by UV/H2O2 and UV/PDS. *Water Res.* **2018**, *145*, 487–497. [CrossRef] [PubMed]
21. Miklos, D.B.; Hartl, R.; Michel, P.; Linden, K.G.; Drewes, J.E.; Hübner, U. UV/H2O2 process stability and pilot-scale validation for trace organic chemical removal from wastewater treatment plant effluents. *Water Res.* **2018**, *136*, 169–179. [CrossRef]
22. Miklos, D.B.; Remy, C.; Jekel, M.; Linden, K.G.; Drewes, J.E.; Hübner, U. Evaluation of advanced oxidation processes for water and wastewater treatment—A critical review. *Water Res.* **2018**, *139*, 118–131. [CrossRef]
23. Miklos, D.B.; Wang, W.L.; Linden, K.G.; Drewes, J.E.; Hübner, U. Comparison of UV-AOPs (UV/H_2O_2, UV/PDS and UV/Chlorine) for TOrC removal from municipal wastewater effluent and optical surrogate model evaluation. *Chem. Eng. J.* **2019**, *362*, 537–547. [CrossRef]
24. Guo, K.; Wu, Z.; Yan, S.; Yao, B.; Song, W.; Hua, Z.; Zhang, X.; Kong, X.; Li, X.; Fang, J. Comparison of the UV/chlorine and UV/H2O2 processes in the degradation of PPCPs in simulated drinking water and wastewater: Kinetics, radical mechanism and energy requirements. *Water Res.* **2018**, *147*, 184–194. [CrossRef]
25. Ter Laak, T.; Tolkamp, H.; Hofman, J. *Geneesmiddelen in de Watercyclus in Limburg*; KWR Watercycle Research Institute: Nieuwegein, The Netherlands, 2013. (In Dutch)
26. Huber, S.A.; Balz, A.; Abert, M.; Pronk, W. Characterisation of aquatic humic and non-humic matter with size-exclusion chromatography—Organic carbon detection—Organic nitrogen detection (LC-OCD-OND). *Water Res.* **2011**, *45*, 879–885. [CrossRef] [PubMed]
27. Hofman-Caris, C.H.M.; Harmsen, D.J.H.; Van Remmen, A.M.; Knol, A.H.; Van Pol, W.L.C.; Wols, B.A. Optimization of UV/H_2O_2 processes for the removal of organic micropollutants from drinking water: Effect of reactor geometry and water pretreatment on EEO values. *Water Sci. Technol.* **2017**, *17*, 508–518. [CrossRef]
28. Wols, B.A.; Hofman-Caris, C.H.M.; Harmsen, D.J.H.; Beerendonk, E.F. Degradation of 40 selected pharmaceuticals by UV/H_2O_2. *Water Res.* **2013**, *47*, 5876–5888. [CrossRef]
29. Hofman-Caris, C.H.M.; Siegers, W.G.; van de Merlen, K.; de Man, A.W.A.; Hofman, J.A.M.H. Removal of pharmaceuticals from WWTP effluent: Removal of EfOM followed by advanced oxidation. *Chem. Eng. J.* **2017**, *327*, 514–521. [CrossRef]
30. Wols, B.A.; Harmsen, D.J.H.; van Remmen, T.; Beerendonk, E.F.; Hofman-Caris, C.H.M. Design aspects of UV/H_2O_2 reactors. *Chem. Eng. Sci.* **2015**, *137*, 712–721. [CrossRef]
31. Vergouwen, A.A.; Pieters, B.J.; Kools, S. *ZORG Inventarisatie van Emissie van Geneesmiddelen uit Zorginstellingen*; Part C; Stichting Toegepast Onderzoek Waterbeheer (STOWA): Amersfoort, The Netherlands, 2011. (In Dutch)
32. Niemuth, N.J.; Jordan, R.; Crago, J.; Blanksma, C.; Johnson, R.; Klaper, R.D. Metformin exposure at environmentally relevant concentrations causes potential endocrine disruption in adult male fish. *Environ. Toxicol. Chem.* **2015**, *34*, 291–296. [CrossRef] [PubMed]
33. Niemuth, N.J.; Klaper, R.D. Emerging wastewater contaminant metformin causes intersex and reduced fecundity in fish. *Chemosphere* **2015**, *135*, 38–45. [CrossRef]
34. Société Publique de Gestion de l'Eau. *Plan Dássainissement par Sous-Bassin Hydrographique (PASH)*; Société Publique de Gestion de l'Eau: Verviers, Belgium, 2005. (In French)
35. Ter Laak, T.L.; Kooij, P.J.F.; Tolkamp, H.; Hofman, J. Different compositions of pharmaceuticals in Dutch and Belgian rivers explained by consumption patterns and treatment efficiency. *Environ. Sci. Pollut. Res.* **2014**, *21*, 12843–12855. [CrossRef] [PubMed]
36. Bourgin, M.; Beck, B.; Boehler, M.; Borowska, E.; Fleiner, J.; Salhi, E.; Teichler, R.; von Gunten, U.; Siegrist, H.; McArdell, C.S. Evaluation of a full-scale wastewater treatment plant upgraded with ozonation and biological post-treatments: Abatement of micropollutants, formation of transformation products and oxidation by-products. *Water Res.* **2018**, *129*, 486–498. [CrossRef]

37. Coppens, L.J.C.; van Gils, J.A.G.; ter Laak, T.L.; Raterman, B.W.; van Wezel, A.P. Towards spatially smart abatement of human pharmaceuticals in surface waters: Defining impact of sewage treatment plants on susceptible functions. *Water Res.* **2015**, *81*, 356–365. [CrossRef]
38. Kümmerer, K. Chapter 7 Benign by design. In *Green and Sustainable Medicinal Chemistry: Methods, Tools and Strategies for the 21st Century Pharmaceutical Industry*; Somerton, L., Sneddon, H.F., Jones, L.C., Clark, J.H., Eds.; RSC Green Chemistry: Cambridge, UK, 2016; pp. 73–81. [CrossRef]

Article

Utilization of Non-Living Microalgae Biomass from Two Different Strains for the Adsorptive Removal of Diclofenac from Water

Ricardo N. Coimbra [1], Carla Escapa [1], Nadyr C. Vázquez [1], Guillermo Noriega-Hevia [2] and Marta Otero [3,*]

1 Department of Applied Chemistry and Physics, IMARENABIO-Institute of Environment, Natural Resources and Biodiversity, Campus de Vegazana s/n, Universidad de León, 24071 León, Spain; ricardo.decoimbra@unileon.es (R.N.C.); carla.escapa@unileon.es (C.E.); nvazqs00@estudiantes.unileon.es (N.C.V.)

2 CALAGUA—Unidad Mixta UV-UPV, IIAMA-Research Institute of Water and Environmental Engineering, Camí de Vera s/n, Universitat Politècnica de València, 46022 València, Spain; guillermonoriegahevia@gmail.com

3 CESAM-Centre for Environmental and Marine Studies, Department of Environment and Planning, Campus Universitário de Santiago, University of Aveiro, 3810-193 Aveiro, Portugal

* Correspondence: marta.otero@ua.pt; Tel.: +351-234247094

Received: 3 September 2018; Accepted: 6 October 2018; Published: 9 October 2018

Abstract: In the present work, the adsorptive removal of diclofenac from water by biosorption onto non-living microalgae biomass was assessed. Kinetic and equilibrium experiments were carried out using biomass of two different microalgae strains, namely *Synechocystis* sp. and *Scenedesmus* sp. Also, for comparison purposes, a commercial activated carbon was used under identical experimental conditions. The kinetics of the diclofenac adsorption fitted the pseudo-second order equation, and the corresponding kinetic constants indicating that adsorption was faster onto microalgae biomass than onto the activated carbon. Regarding the equilibrium results, which mostly fitted the Langmuir isotherm model, these pointed to significant differences between the adsorbent materials. The Langmuir maximum capacity (Q_{max}) of the activated carbon (232 mg·g^{-1}) was higher than that of *Scenedesmus* sp. (28 mg·g^{-1}) and of *Synechocystis* sp. (20 mg·g^{-1}). In any case, the Q_{max} values determined here were within the values published in the recent scientific literature on the utilization of different adsorbents for the removal of diclofenac from water. Still, *Synechocystis* sp. showed the largest K_L fitted values, which points to the affinity of this strain for diclofenac at relative low equilibrium concentrations in solution. Overall, the results obtained point to the possible utilization of microalgae biomass waste in the treatment of water, namely for the adsorption of pharmaceuticals.

Keywords: emerging contaminants (ECs); sorption; wastewater treatment; bioremediation; algae

1. Introduction

Microalgae are photosynthetic microorganisms capable of using CO_2 as a carbon source. Thus, as the accumulation of CO_2 in the atmosphere is one of the most serious environmental issues to be faced nowadays, the possibility of using microalgae for its sequestration has received great attention [1]. Still, the implementation of CO_2 sequestration by microalgae is mostly limited by techno-economic constrains [2]. An option to increase the cost-effectiveness is the cultivation of microalgae in wastewater, which is a complex mixture that may serve as a source of nutrients and water [3]. This strategy allows for nutrient recycling with savings in microalgae cultivation costs and, simultaneously contributes to enhancing the sustainability of wastewater treatment [4,5].

Interest in microalgae-based wastewater treatment has increased in recent years since, while growing, these microorganisms are able to uptake pollutants like nutrients [6] and trace metals [7], but also emerging contaminants (ECs) such as pharmaceuticals [8–10]. The latter represent an especially worrying class of contaminants since they were designed to provoke a physiological response and their presence in the aquatic environment may affect non-target individuals. Among the different treatments proposed for the removal of pharmaceuticals from wastewater, microalgae-based systems have been proved to be effective either in close [8] or open [11] systems. Whatever the system configuration, biodegradation, together with bioadsorption and bioaccumulation, have been indicated as the main mechanisms for the removal of pharmaceuticals from wastewater [5].

Comparatively with research on the uptake of pharmaceuticals by growing microalgae in wastewater, the utilization of non-living microalgae biomass for the adsorptive removal of these pollutants is still in its early stages [12,13]. That is not the case of the well-known adsorption capacity of microalgae to remove other pollutants such as metals [14,15] or dyes [16]. Still, in the case of pharmaceuticals, a main advantage of the application of adsorption processes for their removal is that transformation products, which may be generated during treatments involving degradation [17,18], are not produced. On the other hand, the utilization of the residual microalgae biomass for the adsorption of pollutants from water following the extraction of lipids, has been pointed to as a feasible zero-waste strategy to improve the sustainability of microalgae cultivation [13].

In this context, the aim of this work was to study the adsorptive removal of diclofenac by non-living microalgae biomass of two different strains, namely *Scenedesmus* sp. (Chlorophyceae) and *Synechocystis* sp. (Cyanophyceae). For comparison purposes, a commercial activated carbon was used as a reference under the same experimental conditions as microalgae biomass. Diclofenac was selected as target pharmaceutical since it is a nonsteroidal anti-inflammatory drug (NSAID), it is widely consumed, it is one of the pharmaceuticals most frequently present in effluents from sewage treatment plants [19], and it is potentially toxic towards several organisms such as fish and mussels [20]. Moreover, concern about the presence of diclofenac in the aquatic environment has led to its inclusion in the first watch list (EU Decision 2015/495) to support future revisions of the list of priority substances within the Water Framework Directive (2000/60/EC) (WFD) [21].

2. Materials and Methods

2.1. Microalgae and Culture Conditions

Microalgae from two different genera were used in this work: (i) *Scenedesmus* sp. (SAG 276-1), which was purchased from the *Sammlung von Algenkulturen der Universität Göttingen* (Culture Collection of Algae at Göttingen University, international acronym SAG); and (ii) *Synechocystis* sp., which was isolated from natural freshwater in the surroundings of the province of León [22]. It is to note that the term microalgae was here used in a wide sense, since Cyanophyceae (commonly known as blue green algae) have prokaryotic cell structure like bacteria and, because of that, have also been named as cyanobacteria. An inoculum of each strain was maintained in Erlenmeyer flasks (250 mL) containing the standard medium Mann and Myers [23] and kept inside a vegetal culture chamber under controlled growth conditions: temperature (25 ± 1 °C), irradiance (175 μmol photons $m^{-2}{\cdot}s^{-1}$), photoperiod (12:12) and shaking (250 rpm). Then, the cultures were grown in bubbling column photobioreactors (PBRs) with an operation volume of 9 L. PBRs were kept in vegetal culture chambers under controlled conditions, namely at 27–30 °C, 16:8 photoperiod of light:darkness, and irradiance of 650 $\mu E{\cdot}m^{-2}{\cdot}s^{-1}$. The microalgae cultures were aerated with filtered air (0.22 μm sterile filters, Millex FG50 Millipore (Merck Millipore, Burlington, MA, USA)) at 0.3 v/v/min. Air was enriched with CO_2 at 7% v/v, which was injected on demand to keep a constant pH (pH = 7.5 ± 0.5), as controlled by a pH sensor.

2.2. Adsorbent Materials and Adsorption Experiments

For the two different strains, the cellular suspension from each of the aforementioned cultures was centrifuged (7800 rpm, 7 min) to separate microalgae biomass from the culture medium. Then, the biomass was washed twice with distilled water, frozen and lyophilized. Before its use as a biosorbent, the lyophilized biomass was grinded and homogenized. For comparison purposes, a commercial activated carbon (PULSORB WP260 (Chemviron Carbon, Feluy, Belgium)), which was generously provided by Chemviron Carbon, was used in this work.

Diclofenac sodium ($C_{14}H_{10}Cl_2NNaO_2$, ≥99%) (Sigma-Aldrich, Madrid, Spain) was used in the adsorption experiments. The concentration of diclofenac in liquid phase was analyzed by a Waters HPLC 600 equipped with a 2487 Dual λ Absorbance Detector (Waters Corporation, Milford, MA, USA), a Phenomenex C18 column (Phenomenex España S.L.U., Madrid, Spain), (5 μm, 110 Å, 250 × 4.6 mm), a Rheodyne injector (Waters Corporation, Milford, MA, USA), and a 50 μL loop (Waters Corporation, Milford, MA, USA). The detection wavelength was 276.5 nm and the mobile phase consisted of acetonitrile:water:orthophosphoric acid (70:30:0.1, v/v/v), which was pumped at 1 mL·min^{-1}. For the mobile phase preparation, HPLC quality acetonitrile (CH_3CN) from LAB-SCAN, orthophosphoric acid (H_3PO_4) from Panreac and ultrapure water obtained by a Millipore System were used. Before use, the mobile phase mixture was passed through a Millipore filter (0.45 μm) and degassed by ultrasound application during 30 min. On the other hand, all the samples were centrifuged at 7500 rpm for 10 min (SIGMA 2-16P centrifuge (Sigma Laborzentrifugen GmbH, Osterode am Harz, Germany), before analysis.

The adsorption experiments were carried out under stirring and batch operation following a parallel approach (a reactor was run by triplicate for each desired time and/or adsorbent mass). Reactors were Erlenmeyer flasks (100 mL) containing a volume (V) of 50 mL of solution with a known initial concentration (C_i) of adsorbate, namely diclofenac, together with a known mass (m_{ads}) of each adsorbent. Since the adsorption behavior of an adsorbent towards a certain adsorbate is not known a priori, preliminary test were here settled at different C_i and m_{ads} for each material. These tests aimed at the selection of appropriate asorbent to adsorbate ratios for the subsequent kinetic and equilibrium experiments. The choice of the C_i and the m_{ads} for each material, which are specified in the following sections, was such to ensure: (i) a significant change of the adsorbate concentration in solution through adsorption experiments; and (ii) a final concentration of adsorbate that might be accurately and precisely determined by the analytic methodology used.

2.2.1. Adsorption Kinetics

For each adsorbent, adsorption kinetic experiments were first carried out in order to determine the time necessary to attain adsorption equilibrium (t_e). In each reactor, a diclofenac solution with C_i = 100 mg·L^{-1} was stirred at 250 rpm under controlled temperature (25 ± 1 °C) together with a known m_{ads}. In the case of *Scenedesmus* sp. and *Synechocystis* sp., 0.05 g of biomass were employed whereas 0.005 g of activated carbon were used in kinetic experiments. After stirring during the desired time (t), reactors were withdrawn, and a sample of the liquid phase was analyzed for the residual concentration of diclofenac (C_t). Three replicated reactors were run for each considered adsorbent and time. Furthermore, blanks (adsorbent + distilled water, without diclofenac in the aqueous phase) and controls (diclofenac solution with no adsorbent) were also run in triplicate. Throughout experiments, the pH of the solutions was not fixed at any initial value neither buffered, but stability in the values was observed along the kinetic experiments (7.0 ± 0.5).

At each t, the adsorbed concentration of diclofenac onto each adsorbent (q_t) was determined by a mass balance, as indicated by Equation (1):

$$q_t = \frac{(C_i - C_t)}{m_{ads}} \times V \quad (1)$$

Fittings of the experimental results to the pseudo-first and pseudo second-order kinetic equations are shown together with the experimental results in Figure 1. The kinetic parameters derived from these fittings are depicted in Table 1. Fittings to both equations were reasonably good, with $r^2 > 0.98$, with the pseudo-second kinetic equation describing results slightly better. Both the kinetic models here considered are based on the adsorbed concentration at the equilibrium (q_e). However, as can be seen in Figure 1, the pseudo-first order model is valid just at the initial stage of adsorption while the pseudo-second model provides good fitting over the whole time range. Hence, in the case of the k_1, values determined for the three materials were not significantly different, which points to the fact that the initial uptake of diclofenac adsorption by the activated carbon and the microalgae biomasses showed a similar rate. Then, differences in the kinetics occurred at a second stage, which was evidenced by the fitted values of the k_2 rate constants. These k_2 were equal for both microalgae strains and larger than that of activated carbon, which indicated that, on the whole, the adsorption kinetic was comparatively faster onto microalgae biomass.

Table 1. Parameters from the experimental results fittings to the kinetic (pseudo-first order kinetic equation and pseudo-second order equation) and equilibrium isotherm (Langmuir and Freundlich equilibrium isotherms) models considered.

Model	Parameter	*Scenedesmus* sp.	*Synechocystis* sp.	Activated Carbon
		Kinetic Equations		
Pseudo-first order	k_1 (min^{-1})	0.0388 ± 0.0041	0.0393 ± 0.0024	0.0375 ± 0.0021
	q_e (mg·g^{-1})	20.19 ± 0.54	17.55 ± 0.30	184.90 ± 2.98
	r^2	0.981	0.9944	0.9951
	$S_{y.x}$	1.05	0.52	5.16
Pseudo-second order	k_2 (g·m^{-1}·min^{-1})	0.0023 ± 0.0002	0.0025 ± 0.0002	0.00022 ± 0.00002
	q_e (mg·g^{-1})	22.64 ± 0.35	19.90 ± 0.34	210.80 ± 4.50
	r^2	0.9964	0.9968	0.9953
	$S_{y.x}$	0.45	0.40	5.09
		Equilibrium Isotherms		
Freundlich	K_F (mg·g^{-1} (mg·L^{-1})$^{-N}$)	3.48 ± 0.17	5.40 ± 1.01	43.55 ± 7.48
	N	2.36 ± 0.07	3.42 ± 0.61	2.80 ± 0.36
	r^2	0.9989	0.9424	0.9579
	$S_{y.x}$	0.26	1.78	15.23
Langmuir	Q_{max} (mg·g^{-1})	28.34 ± 1.19	19.76 ± 0.57	232.20 ± 7.41
	K_L (L·mg^{-1})	0.039 ± 0.005	0.143 ± 0.018	0.076 ± 0.007
	r^2	0.9941	0.9919	0.9932
	$S_{y.x}$	0.57	0.66	6.12

Note: r^2—Correlation coefficient; $S_{y.x}$—Standard error of the regression.

The diclofenac adsorption equilibrium isotherms using *Scenedesmus* sp. biomass, *Synechocystis* sp. biomass and activated carbon as adsorbents are shown in Figure 2.

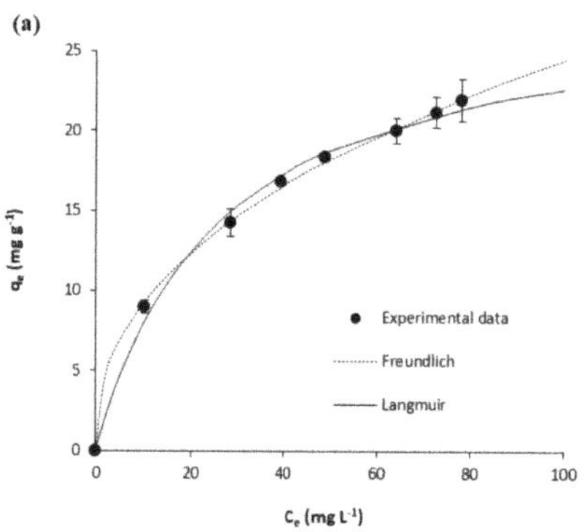

Figure 2. *Cont.*

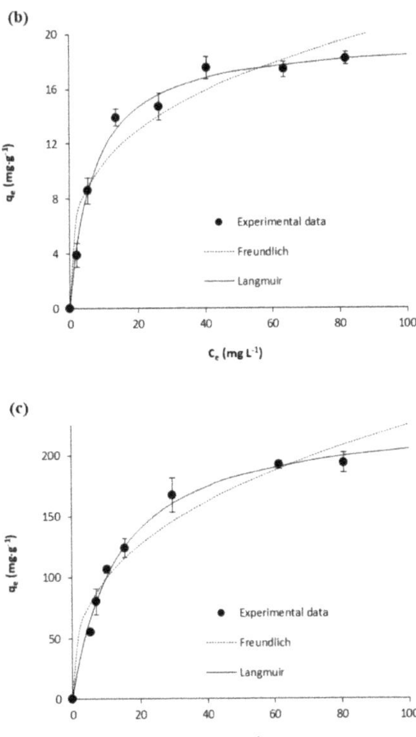

Figure 2. Equilibrium results on the adsorption of diclofenac onto (**a**) *Scenedesmus* sp. biomass; (**b**) *Synechocystis* sp. biomass; and (**c**) activated carbon. Experimental data on the equilibrium adsorbed concentration of diclofenac (q_e, mg·g^{-1}) versus the equilibrium diclofenac concentration in the liquid phase (C_e, mg·L^{-1}) are represented together with fittings to the Langmuir and Freundlich equilibrium isotherm models. Notes: error bars stand for standard deviation (N = 3). The scale of the axis has been adjusted for a better visualization of results.

Fittings of equilibrium experimental results to the Freundlich and Langmuir models are represented in Figure 2, the corresponding fitted parameters being depicted in Table 1.

In the case of diclofenac adsorption onto *Scenedesmus* sp. biomass, equilibrium results fitted the Freundlich and Langmuir isotherm models, with $r^2 > 0.99$ in both cases. However, for both the *Synechocystis* sp. biomass and the commercial activated carbon, equilibrium results were better described by the Langmuir isotherm.

Figure 2 makes evident that, at the equilibrium, the adsorptive removal of diclofenac by the activated carbon used here was larger than that of microalgae biomasses. On the other hand, the diclofenac adsorption capacity of *Scenecesmus* sp. was significantly larger than that of *Synechocystis* sp., which may be confirmed by Q_{max} values in Table 1. According to the Langmuir isotherm model [27], the Q_{max}, which is the maximum adsorption capacity, corresponds to the saturation of a monolayer of adsorbate molecules on the adsorbent surface, that is, when all the adsorption sites of the adsorbent are occupied by adsorbate molecules. Therefore, each adsorbent possesses a unique Q_{max} for each adsorbate and, in wastewater treatment applications, a larger value of Q_{max} implies that the adsorbent material will have a longer useful lifetime. Hence, Q_{max} is used for the prediction of the adsorbent performance in real systems and for the design of adsorbers at different scales [28]. In this work, the Q_{max} determined for activated carbon (232 mg·g^{-1}) was larger than that of *Scenedesmus* sp. and

Synechocystis sp. (28 mg·g^{-1} and 20 mg·g^{-1}, respectively). In any case, the here obtained Q_{max} values for the adsorption of diclofenac onto microalgae biomass are higher than those determined for the adsorption of different polyphenols (8 mg·g^{-1} < Q_{max} < 19 mg·g^{-1}) onto non-living *Chlorella* sp. biomass [29] but lower than for the adsorption of acetaminophen onto *Synechocystis* sp. (52 mg·g^{-1}). With respect to other materials used for the adsorptive removal of diclofenac from water, Table 2 shows recently Q_{max} published values for adsorbents of different nature. As may be seen, the range is quite large and comprises the here obtained Q_{max}.

Regarding the K_L, which points to the affinity of an adsorbent towards the adsorbate, the fitted value determined for *Synechocystis* sp. (0.14 L·mg^{-1}) is within values obtained for the adsorption of polyphenols onto *Chlorella* sp. (0.09–0.022 L·mg^{-1}) [29]. For the commercial activated carbon and *Scenedesmus* sp., the K_L determined was one order of magnitude lower than that of *Synechocystis* sp., as for the steeper isotherm of the latter (Figure 2). Therefore, although *Synechocystis* sp. displayed the smallest value of maximum adsorption capacity, this Q_{max} was attained at relatively low C_e of diclofenac in solution.

Table 2. Maximum adsorption capacities Q_{max} (mg·g^{-1}) of different types of adsorbents used for the adsorptive removal of diclofenac from water (single non-competitive adsorption; T: 25 ± 2 °C; pH: 7 ± 2).

Adsorbent	Q_{max} (mg·g^{-1})	Reference
Activated onion skin	134	[30]
Metal azolate framework-6	503	[31]
Activated cork	79	[31]
Pyrolyzed pulp mill sludge	27	[32]
Granular activated carbon	36	[33]
Activated carbon from olive stones	11	[34]
Ionic liquid modified biomass	197	[35]
MIEX® resin	52	[36]
Molecular imprinted polymer	160	[37]
Powder activated carbon	301	[38]
Polymeric resin	39	[38]

To the best of our knowledge, there are not previous records in the literature on the adsorptive different performance of *Scenedesmus* sp. and *Synechocystis* sp. biomass observed in this work. It must be highlighted that, in the present work, microalgae biomass used was not previously modified neither subjected to thermal treatment. Thus, differences between the two strains regarding the adsorption of diclofenac may be related to their cell wall and biochemical composition. In fact, it has already being pointed out that the microalgae cell surface possesses a rich variety of binding possibilities for a whole range of chemical compounds [29].

Microalgae constitute a group of microorganisms that are easy to culture due to their high growth rates and productivities and, therefore, microalgae biotechnological applications are under expansion [29]. Among the strategies to reduce costs associated with the culture of microalgae is the utilization of flue gases as CO_2 supply and wastewater as nutrients and freshwater source [3]. In this way, microalgae could be used for the biosequestration of CO_2 while accomplishing wastewater treatment [39]. In any case, during cultivation, waste microalgae biomass is generated and a use should be given to this biomass within the actual circular economy context. Therefore, the utilization of microalgae biomass as adsorbent may be an option for increasing the sustainability of microalgae culture. Furthermore, such a use is especially interesting since it may be implemented after lipid extraction from non-living microalgae [13]. As diclofenac is among the pharmaceuticals within the first watch list in the European Union (EU) [21], the novel results obtained in this work on its uptake by non-living microalgae biomass point to the possible application of this biomass for the adsorptive removal of this sort of emerging contaminant. Promissory results obtained in the present work show that this is a new line of research that is worth to further exploring. In this sense, future studies are to be done on the application at real systems, in which fixed-bed microalgae adsorbers may be implemented by the immobilization of microalgae biomass. Although there are no published results for the removal of pharmaceuticals, Saeed and Iqbal [40] immobilized a blue green microalga, namely *Synechococcus* sp. on loofa (*Luffa cylindrical*) sponge for the fixed-bed adsorptive removal of cadmium from water, a strategy that was later adopted by Chen et al. [41], who used *Scenedesmus obliquus* as biosorbent.

4. Conclusions

The microalgae non-living biomass of two different strains, namely *Scenedesmus* sp. and *Synechocystis* sp. was used for the adsorptive removal of diclofenac from water. Kinetic and equilibrium results were compared with those obtained by a commercial activated carbon under identical experimental conditions. Fittings of the kinetic experimental results to the pseudo-second kinetic equation showed that the rate of diclofenac uptake from aqueous solution was similar for both microalgae strains and faster than that of activated carbon. Regarding the equilibrium experimental results, the Langmuir isotherm model described the results for the three adsorbents. The fitted values of the Langmuir maximum adsorption capacity (Q_{max}) were 232, 28 and 20 mg·g^{-1} of diclofenac onto the activated carbon, *Scenedesmus* sp. biomass and *Synechocystis* sp. biomass, respectively. These values are within recently published Q_{max} for the adsorptive removal of diclofenac from water using different adsorbents. Differently from these adsorbents in the literature, microalgae biomass here used was neither modified nor treated, its use as biosorbent being an option to explore in view of a sustainable zero-waste strategy for the culture of microalgae.

Author Contributions: R.N.C., C.E. and M.O. conceived the work and the experimental design; R.N.C., C.E., N.C.V., G.N.-H. and M.O. performed the experiments and chemical analysis; R.N.C., C.E. and M.O. analysed the results and wrote the manuscript. All the authors approved the submitted final version.

Funding: This research was funded by Fundação para a Ciência e a Tecnologia (FCT, Lisboa, Portugal), grant number IF/00314/2015; by the Ministry of Economy and Competitiveness, State Secretariat for Research, Development and Innovation (Madrid, Spain), grant number RYC 2010-05634; and by University of León (Spain), grant number UXXI2016/00128. C.E. was funded by the Ministry of Education, Culture and Sports (Madrid, Spain), grant number FPU12/03073.

Acknowledgments: Thanks are due for financial support to CESAM (UID/AMB/50017—POCI-01-0145-FEDER-007638), to FCT/MCTES through national funds (PIDDAC), and co-funding by FEDER, within the PT2020 Partnership Agreement and Compete 2020.

Conflicts of Interest: The authors declare the inexistence of conflict of interest. They also declare that the founding agents did not participate in nor decide on the design of the study, the chemical analysis, the interpretation of results, and the writing and publishing of the manuscript.

References

1. Cheah, W.Y.; Show, P.L.; Chang, J.S.; Ling, T.C.; Juan, J.C. Biosequestration of atmospheric CO_2 and flue gas-containing CO_2 by microalgae. *Bioresour. Technol.* **2015**, *184*, 190–201. [CrossRef] [PubMed]
2. Kassim, M.A.; Meng, T.K. Carbon dioxide (CO_2) biofixation by microalgae and its potential for biorefinery and biofuel production. *Sci. Total Environ.* **2017**, *584–585*, 1121–1129. [CrossRef] [PubMed]
3. Acién, F.G.; Fernández, J.M.; Magan, J.J.; Molina, E. Production cost of a real microalgae production plant and strategies to reduce it. *Biotechnol. Adv.* **2012**, *30*, 1344–1353. [CrossRef] [PubMed]
4. Kim, J.-Y.; Kim, H.-W. Photoautotrophic microalgae screening for tertiary treatment of livestock wastewater and bioresource recovery. *Water* **2017**, *9*, 192. [CrossRef]
5. Xiong, J.-Q.; Kurade, M.B.; Jeon, B.-H. Can microalgae remove pharmaceutical contaminants from water? *Trends Biotechnol.* **2018**, *36*, 30–44. [CrossRef] [PubMed]
6. Khiewwijit, R.; Rijnaarts, H.; Temmink, H.; Keesman, K.J. Glocal assessment of integrated wastewater treatment and recovery concepts using partial nitritation/Anammox and microalgae for environmental impacts. *Sci. Total Environ.* **2018**, *628*, 74–84. [CrossRef] [PubMed]
7. Rugnini, L.; Costa, G.; Congestri, R.; Bruno, L. Testing of two different strains of green microalgae for Cu and Ni removal from aqueous media. *Sci. Total Environ.* **2017**, *601–602*, 959–967. [CrossRef] [PubMed]
8. Hom-Diaz, A.; Jaén-Gil, A.; Bello-Laserna, I.; Rodríguez-Mozaz, F.; Vicent, T.; Barceló, D.; Blánquez, P. Performance of a microalgal photobioreactor treating toilet wastewater: Pharmaceutically active compound removal and biomass harvesting. *Sci. Total Environ.* **2017**, *592*, 1–11. [CrossRef] [PubMed]

9. Escapa, C.; Coimbra, R.N.; Paniagua, S.; García, A.I.; Otero, M. Comparative assessment of pharmaceutical removal from wastewater by the microalgae Chlorella sorokiniana, Chlorella vulgaris and Scenedesmus obliquus. In *Biological Wastewater Treatment and Resource Recovery*; IntechOpen: London, UK, 2017.
10. Escapa, C.; Coimbra, R.N.; Paniagua, S.; García, A.I.; Otero, M. Comparative assessment of diclofenac removal from water by different microalgae strains. *Algal Res.* **2016**, *18*, 127–134. [CrossRef]
11. de Godos, I.; Muñoz, R.; Guieysse, B. Tetracycline removal during wastewater treatment in high-rate algal ponds. *J. Hazard. Mater.* **2012**, *229*, 446–449. [CrossRef] [PubMed]
12. Escapa, C.; Coimbra, R.N.; Nuevo, C.; Vega, S.; Paniagua, S.; García, A.I.; Calvo, L.F.; Otero, M. Valorization of microalgae biomass by its use for the removal of paracetamol from contaminated water. *Water* **2017**, *9*, 312. [CrossRef]
13. Angulo, E.; Bula, L.; Mercado, I.; Montaño, A.; Cubillán, N. Bioremediation of Cephalexin with non-living *Chlorella* sp. biomass after lipid extraction. *Bioresour. Technol.* **2018**, *257*, 17–22. [CrossRef] [PubMed]
14. Brinza, L.; Dring, M.J.; Gavrilescu, M. Marine micro and macro algal species as biosorbents for heavy metals. *Env. Eng. Manag. J.* **2007**, *6*, 237–251. [CrossRef]
15. Sutkowy, M.; Kłosowski, G. Use of the coenobial green algae *Pseudopediastrum boryanum* (Chlorophyceae) to remove hexavalent chromium from contaminated aquatic ecosystems and industrial wastewaters. *Water* **2018**, *10*, 712. [CrossRef]
16. Tsai, W.-T.; Chen, H.-R. Removal of malachite green from aqueous solution using low-cost chlorella-based biomass. *J. Hazard. Mater.* **2010**, *175*, 844–849. [CrossRef] [PubMed]
17. Peng, F.; Ying, G.; Yang, B.; Liu, S.; Lai, H.; Liu, Y.; Chen, Z.; Zhou, G. Biotransformation of progesterone and norgestrel by two freshwater microalgae (*Scenedesmus obliquus* and *Chlorella pyrenoidosa*): transformation kinetics and products identification. *Chemosphere* **2014**, *95*, 581–588. [CrossRef] [PubMed]
18. Escapa, C.; Torres, T.; Neuparth, T.; Coimbra, R.N.; García, A.I.; Santos, M.M.; Otero, M. Zebrafish embryo bioassays for a comprehensive evaluation of microalgal efficiency in the removal of diclofenac from water. *Sci. Total Environ.* **2018**, *640*, 1024–1033. [CrossRef] [PubMed]
19. Pal, A.; Gin, K.Y.H.; Lin, A.Y.-C.; Reinhard, M. Impacts of emerging organic contaminants on freshwater resources: Review of recent occurrences, sources, fate and effects. *Sci. Total Environ.* **2010**, *408*, 6062–6069. [CrossRef] [PubMed]
20. Lonappan, L.; Brar, S.K.; Das, R.K.; Verma, M.; Surampalli, R.Y. Diclofenac and its transformation products: Environmental occurrence and toxicity—A review. *Environ. Int.* **2016**, *96*, 127–138. [CrossRef] [PubMed]
21. Barbosa, M.O.; Moreira, N.F.F.; Ribeiro, A.R.; Pereira, M.F.R.; Silva, A.M.T. Occurrence and removal of organic micropollutants: An overview of the watch list of EU Decision 2015/495. *Water Res.* **2016**, *94*, 257–279. [CrossRef] [PubMed]
22. Martínez, L.; Otero, M.; Morán, A.; García, A.I. Selection of native freshwater microalgae and cyanobacteria for CO_2 biofixation. *Environ. Technol.* **2013**, *34*, 3137–31435. [CrossRef] [PubMed]
23. Mann, J.; Myers, J. On pigments growth and photosynthesis of *Phaeodactylum Tricornutum*. *J. Phycol.* **1968**, *4*, 349–355. [CrossRef] [PubMed]
24. Lagergren, S. Zur theorie der sogenannten adsorption gelöster stoffe. Kungliga Svenska Vetenskapsakademiens. *Handlingar* **1808**, *24*, 1–39.
25. Ho, I.S.; McKay, G. Pseudo-second order model for sorption processes. *Process Biochem.* **2011**, *34*, 451–465. [CrossRef]
26. Freundlich, H. Über die Adsorption in Lösungen. *Z. Phys. Chem.* **1906**, *57*, 385–470. [CrossRef]
27. Langmuir, I. The Adsorption of Gases on Plane Surfaces of Glass, Mica and Platinum. *J. Am. Chem. Soc.* **1918**, *40*, 1361–1403. [CrossRef]
28. Otero, M.; Zabkova, M.; Grande, C.A.; Rodrigues, A.E. Fixed-bed adsorption of salicylic acid onto polymeric adsorbents and activated charcoal. *Ind. Eng. Chem. Res.* **2005**, *44*, 927–936. [CrossRef]
29. Jelínek, L.; Procházková, G.; Quintelas, C.; Beldíková, E.; Brányik, T. *Chlorella vulgaris* biomass enriched by biosorption of polyphenols. *Algal Res.* **2015**, *10*, 1–7. [CrossRef]
30. Abbas, G.; Javed, I.; Iqbal, M.; Haider, R.; Hussain, F.; Qureshi, N. Adsorption of non-steroidal anti-inflammatory drugs (diclofenac and ibuprofen) from aqueous medium onto activated onion skin. *Desalin. Water Treat.* **2017**, *95*, 274–285. [CrossRef]

31. An, H.J.; Bhadra, B.N.; Khan, N.A.; Jhung, S.H. Adsorptive removal of wide range of pharmaceutical and personal care products from water by using metal azolate framework-6-derived porous carbon. *Chem. Eng. J.* **2018**, *343*, 447–454. [CrossRef]
32. Coimbra, R.N.; Calisto, V.; Ferreira, C.I.A.; Esteves, V.I.; Otero, M. Removal of pharmaceuticals from municipal wastewater by adsorption onto pyrolyzed pulp mill sludge. *Arab. J. Chem.* **2015**. [CrossRef]
33. de Franco, M.A.E.; de Carvalho, C.B.; Bonetto, M.M.; de Pelegrini Soares, R.; Féris, L.A. Diclofenac removal from water by adsorption using activated carbon in batch mode and fixed-bed column: Isotherms, thermodynamic study and breakthrough curves modelling. *J. Clean. Prod.* **2018**, *181*, 145–154. [CrossRef]
34. Larous, S.; Meniai, A.-H. Adsorption of diclofenac from aqueous solution using activated carbon prepared from olive stones. *Int. J. Hydrogen Energy* **2016**, *41*, 10380–10390. [CrossRef]
35. Lawal, I.A.; Moodley, B. Sorption mechanism of pharmaceuticals from aqueous medium on ionic liquid modified biomass. *J. Chem. Tech. Biotech.* **2017**, *92*, 808–818. [CrossRef]
36. Lu, X.; Shao, Y.; Gao, N.; Chen, J.; Zhang, Y.; Wang, Q.; Lu, Y. Adsorption and removal of clofibric acid and diclofenac from water with MIEX resin. *Chemosphere* **2016**, *161*, 400–411. [CrossRef] [PubMed]
37. Samah, N.A.; Sánchez-Martín, M.-J.; Sebastián, R.M.; Valiente, M.; López-Mesas, M. Molecularly imprinted polymer for the removal of diclofenac from water: Synthesis and characterization. *Sci. Total Environ.* **2018**, *631*, 1534–1543. [CrossRef] [PubMed]
38. Coimbra, R.N.; Escapa, C.; Paniagua, S.; Otero, M. Adsorptive removal of diclofenac from ultrapure and wastewater: a comparative assessment on the performance of a polymeric resin and activated carbons. *Desalin. Water Treat.* **2016**, *57*, 27914–27923. [CrossRef]
39. Markou, G.; Wang, L.; Ye, J.; Unc, A. Using agro-industrial wastes for the cultivation of microalgae and duckweeds: Contamination risks and biomass safety concerns. *Biotechnol. Adv.* **2018**, *36*, 1238–1254. [CrossRef] [PubMed]
40. Chen, B.-Y.; Chen, C.-Y.; Guo, W.-Q.; Chang, H.-W.; Chen, W.-M.; Lee, D.-J.; Huang, C.-C.; Ren, N.-Q.; Chang, J.-S. Fixed-bed biosorption of cadmium using immobilized *Scenedesmus obliquus* CNW-N cells on loofa (*Luffa cylindrica*) sponge. *Bioresource Technol.* **2014**, *160*, 175–181. [CrossRef] [PubMed]
41. Saeed, A.; Iqbal, M. Immobilization of blue green microalgae on loofa sponge to biosorb cadmium in repeated shake flask batch and continuous flow fixed bed column reactor system. *World J. Microbiol. Biotechnol.* **2016**, *22*, 775–782. [CrossRef]

Article

Removal of Multi-Class Antibiotic Drugs from Wastewater Using Water-Soluble Protein of *Moringa stenopetala* Seeds

Temesgen Girma Kebede, Simiso Dube * and Mathew Muzi Nindi

Department of Chemistry, University of South Africa, Corner Christian de Wet and Pioneer Avenue, Florida 1709, South Africa; tgkkebede@gmail.com (T.G.K.); nindimm@unisa.ac.za (M.M.N.)

* Correspondence: Dubes@unisa.ac.za; Tel.: +27-11-670-9308

Received: 17 November 2018; Accepted: 17 December 2018; Published: 21 March 2019

Abstract: The removal of ten selected antibiotic drugs belonging to different classes (sulphonamides, fluoroquinolones, macrolides, and tetracycline) was investigated using water-soluble proteins from the seeds of *Moringa stenopetala*. The surface functional groups of water-soluble protein powder before and after removal of antibiotics were characterized using Fourier transform infrared (FTIR). Processing parameters that could affect the removal efficiency, such as initial analyte concentration, protein dosage, and pH were studied. An optimized method was applied to a real wastewater sample collected from Daspoort Wastewater Treatment Plant (WWTP) located in Pretoria, South Africa. Under optimal conditions, the results indicated good agreement between the efficiency of water-soluble proteins to remove antibiotics from the real wastewater sample and from the synthetic wastewater sample prepared in the laboratory using standard solutions with known concentrations. The percentage of removal under optimum conditions (protein dosage of 40 mg, initial analyte concentration of 0.1 mg L^{-1}, and pH 7) was between 85.2 ± 0.01% and 96.3 ± 0.03% for standard mixture solution and from 72.4 ± 0.32% to 92.5 ± 0.84% and 70.4 ± 0.82% to 91.5 ± 0.71% for the real wastewater (effluent and influent) sample.

Keywords: *Moringa stenopetala*; water-soluble proteins; antibiotics removal efficiency

1. Introduction

Antibiotics have saved countless lives since their discovery, and large quantities of these drugs are widely administered and used as antimicrobial drugs throughout the world. Antibiotic drugs are predominantly used to treat bacterial diseases in human therapy and as veterinary medicines to prevent diseases in animal husbandry, and also function as growth promoters, mainly in livestock [1,2]. Excessive usage of antibiotics increases the amount of antibiotic residue discharged into the environment. However, the extent to which antibiotics contaminate the environment has only received attention in recent decades, which could be attributed to the widespread use of and concentration of antibiotics in the aquatic environment, development of advanced and sensitive analytical instruments, and the toxic and chronic effect of antibiotic residues [3]. The reason for the increase in antibiotic concentration levels in the aquatic environment is that they are not completely metabolized in the human/animal body, but rather excreted via urine, animal manure, and/or feces. They are excreted as the parent compounds, metabolites, or water-soluble conjugate compounds and are thus released into the aquatic environment [1,4]. Other sources of antibiotic drugs in the environment include agricultural runoff and the disposal of unused antibiotic drugs from manufacturing industries [5]. As has been reported by different studies, antibiotic drugs have been detected in the influents and effluents of wastewater treatment plants, hospital wastewater, industrial effluent, surface water, groundwater, drinking water, and sediments [1,6–9].

The discharge of antibiotic drugs to the aquatic environment increases the possibility for bacteria to acquire antibiotic resistance genes (ARGs), which are easily transferred to other bacteria through horizontal gene transfer [10]. The development of ARGs in bacteria causes the microbes to become resistant to conventional antibiotic drugs, which had hitherto been effective [11]. The presence of antibiotic resistant bacteria has been observed in wastewater treatment plants, effluents, and surface water in Europe. These drug-resistant bacteria are mainly found in hospital effluents where antibiotics are frequently used.

Amongst the antibiotics drugs, sulphonamides, fluoroquinolones, macrolides, and tetracycline are some of the most frequently detected antibiotic drugs in the aquatic environment [12,13]. These drugs have been detected in municipal wastewater, surface water, ground water, and overland water systems [14,15]. Several recent studies confirm that conventional wastewater treatment plants only partially remove antibiotic drugs from wastewater [6,16].

A number of treatments methods, such as ozonation, chlorination, ultraviolet (UV) irradiation, nanofiltration, reverse osmosis, flocculation, filtration, and adsorption via activated carbons and other materials have been used hitherto. However, most of these methods were developed with the intention of removing heavy metals, hydrophobic drugs, and to treat microbial contaminants rather than pharmaceutically active compounds such as antibiotics [5,17]. Shortcomings have also been identified in some of the methods developed specifically for the purpose of removing antibiotic drugs including photodegradation with UV/catalysts, adsorption by carbon nanotubes, clays, and ion exchange [18–21]. Various materials have been used for the removal of antibiotics, such as zeolite, alumina, silica, mesoporous silica, functionalized mesoporous silica activated carbon, and metal–organic frameworks, biosorbents, agricultural waste, and others [22,23]. It is clearly observed that indeed there is still a need for inexpensive and environmentally friendly yet effective materials for the removal of antibiotics from the various contaminated aquatic environments. In order to reduce the high investments, rigorous research has been carried out to find innovative and cost-effective methods. Amongst the adsorbents, bio-materials have received much attention and have recently become an extensive area of interest due to their effectiveness and versatility in the removal of different kinds of pollutants from the aquatic environment. Moringa has previously been used for water treatment such as flocculation, coagulation, and removal of heavy metals [24,25]. The seeds of Moringa are rich in the water-soluble protein, which has coagulation properties similar to those of alum and synthetic cationic polymers.

The objective of this study was to investigate the removal of multi-class antibiotic drugs, such as sulphonamides, fluoroquinolones, macrolides, and tetracycline, from wastewater by using water-soluble proteins extracted from *Moringa stenopetala* seeds.

2. Materials and Methods

2.1. Chemicals

All standards used were of the highest purity available (≥98%). Sulphanilamide, marbofloxacin, ciprofloxacin, danofloxacin, oxytetracycline, sulphadimethoxine, sulphacetamide, sulphamonomethoxine, sulphamethoxazole, tylosin, and sulphamerazine were obtained from Sigma–Aldrich, (Schnelldorf, Germany). Acetonitrile of high purity HPLC grade (>99.9%) was purchased from Sigma-Aldrich (St. Louis, MO, USA) and a Milli-Q® Integral Water Purification System (Molsheim, France) was used to produce ultra-pure water (18.2 mΩ). The physicochemical properties of the selected compounds are presented in Table 1.

Table 1. Molecular structure and physicochemical properties of selected analytes.

Name of Pharmaceuticals	Structure	pK_{a1} and pK_{a2} (pK_{a3})	Log K_o	Reference
Sulphanilamide	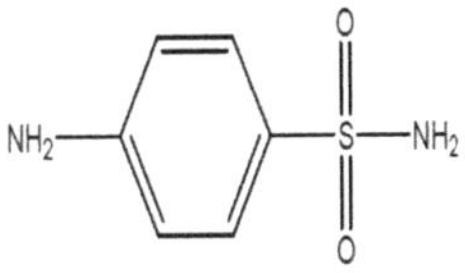	1.94/10.67	−0.62	[26]
Marbofloxacin	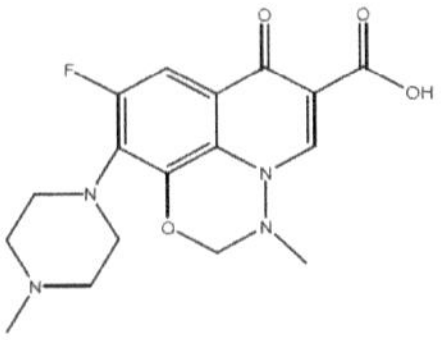	5.77/8.22	0.07 ± 0.01	[27]
Ciprofloxacin	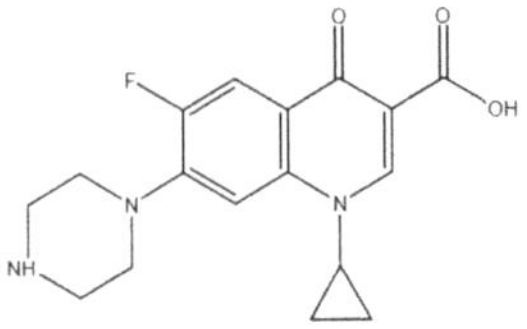	1.84 (0.08)/7.63 (0.01)	0.35	[28,29]
Danofloxacin	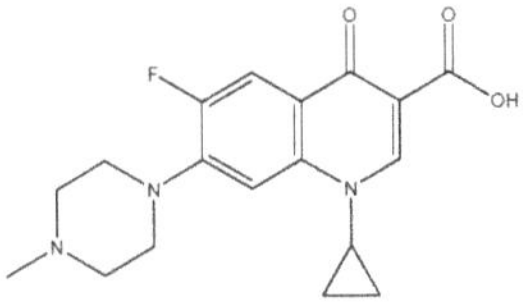	2.06/6.90	0.44	[4]
Oxytetracycline	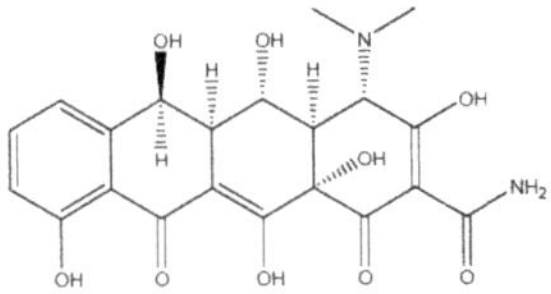	3.3/7.55/9.7	−1.12	[30,31]
Sulphamerazine	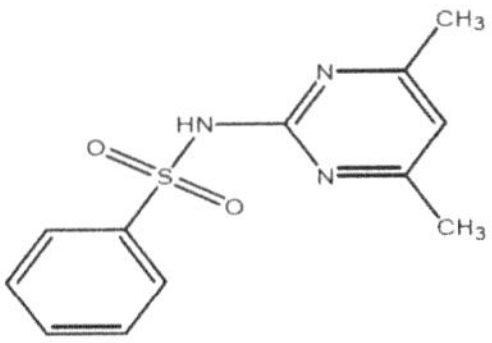	2.06/6.90	0.44	[31]
Sulphamonomethoxine	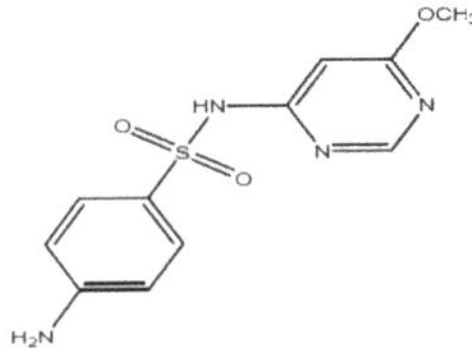	2.07(0.05)/6.91 (0.01)	0.70	[28,32]
Sulphamethoxazole	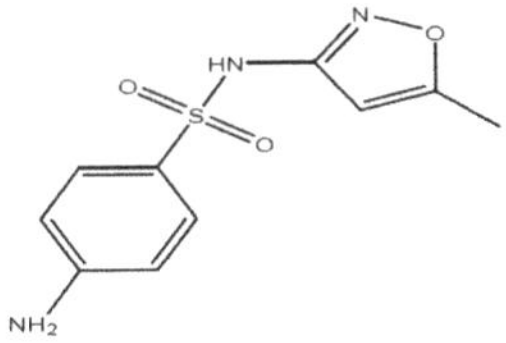	1.6/5.7 (7.40 (0.04))	0.89	[4,28]

Table 1. *Cont.*

Name of Pharmaceuticals	Structure	pK_{a1} and pK_{a2} (pK_{a3})	Log K_o	Reference
Tylosin tartrate		3.31/7.5	-	[31]
Sulphadimethoxine		2.1/6.3	1.63	[32]

2.2. Instrumentation

An Agilent 1220 High-Performance Liquid Chromatography with Diode-Array Detection HPLC-DAD (Agilent Technologies, Waldbronn, Germany) system consisting of a binary high-pressure pump, autosampler, a thermostat column compartment, a fluorescence detector, and refractive index detector was used. ChemStation (version 1.9.0) software (Agilent Technologies, Waldbronn, Germany) was used to process the data. An XTerra MS C18 column (Agilent Technologies, Waldbronn, Germany) was used (4.6 × 100 mm, 3.5 µm), with mobile phase solvent A: 0.1% aqueous formic acid solution, and solvent B: acetonitrile. The column temperature was maintained at 40 °C. The separation was done under gradient elution with the organic phase increasing linearly from 5 to 30% in 7 min and further increasing to 60% within 5 min. The mobile phase was pumped at a flow rate of 1.2 mL·min^{-1} and the detection wavelength was 260 nm, with a post-run time of 1 min before the next injection to equilibrate the column. The chromatographic separation of antibiotic compounds is presented in Figure 1.

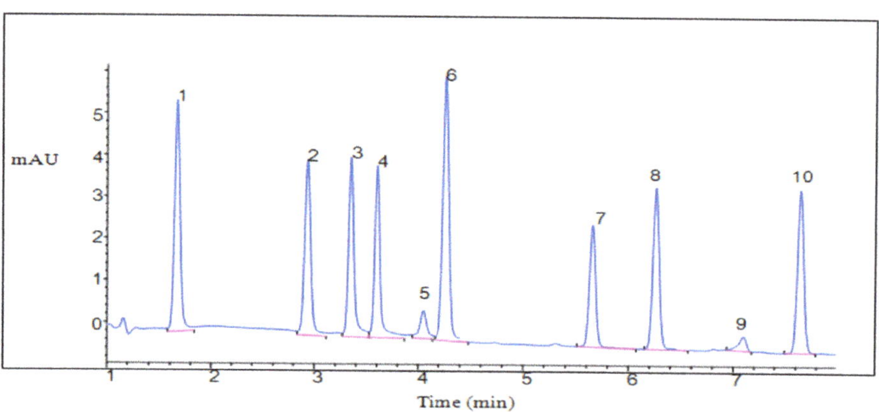

Figure 1. Chromatographic separation of 1 mg L^{-1} standard mixture of ten multiclass antibiotic drugs (sulphanilamide, marbofloxacin, ciprofloxacin, danofloxacin, oxytetracycline, sulphadimethoxine, sulphacetamide, sulphamonomethoxine, sulphamethoxazole, tylosin, and sulphamerazine).

2.3. FTIR Spectroscopy

Fourier transform infrared (FTIR) spectra of freeze-dried, water-soluble proteins before and after antibiotics drug removal were obtained using a Vertex series FTIR spectrophotometer (Bruker Optic GmbH, Hamburg, Germany) with a diamond ATR fitting. The spectra were obtained in transmittance mode with 32 scans at a resolution of 2 cm^{-1} in the 4000–400 cm^{-1}spectral regions. The data were processed using Opus 7.3.139.1294 software (Bruker Optic GmbH, Hamburg, Germany)).

2.4. Preparation of Standard Solutions

All stock solutions for the standards were prepared by weighing 1.5 mg standards and dissolving in acetonitrile/water (50:50) to give a concentration of 1 mg mL^{-1}. All working solutions were prepared by diluting the stock solution in 50:50 (*v*/*v*) acetonitrile and water. The calibration curve was prepared using six concentration levels of standard solutions within the concentration range of 0.01 mg mL^{-1} to 2.5 mg mL^{-1}. All calibration standards were prepared from a stock solution of 1 mg mL^{-1}.

2.5. Extraction of Water-Soluble Proteins

The preparation of the *Moringa stenopetala* seeds for protein extraction was done according to the method described by Kebede et al. [33]. The water-soluble proteins were extracted according to the method adopted from Ndabigengesere et al. [34,35], with a minor modification made as explained in our previous work [33]. Petroleum ether (37% *w*/*v*) was added to the powder and the mixture was stirred for 30 min on a magnetic stirrer to dissolve any fats, oils, or waxes. The undissolved material was then separated via filtration through a Whatman paper No. 1. The residue was dissolved in ultra-high purity water and stirred for 30 min to extract the water-soluble protein. This step was followed by filtration through a Whatman paper No. 3 in order to remove the water-insoluble substances. The filtrate was treated with ammonium sulfate to precipitate proteins from the aqueous extract. The precipitated protein was filtered, re-dissolved in water, and then re-filtered to remove insoluble material. The protein solution was then dialyzed through a cellulose membrane with a molecular cut-off of between 3.5 and 14 kDa. After dialysis, the pure protein was freeze-dried and a white powder was obtained, which was stored at room temperature until ready for use.

2.6. Preparation of Pure Protein for the Removal of Antibiotics

Removal studies were carried out by adding 25 mL aliquots of the standard mixture solution (concentration range 0.1–1.5 mg L^{-1}) in 50 mL Erlenmeyer flasks. The desired amount of protein powder (10–50 mg) was mixed with the standard mixture solution. These solutions were vortexed for 1 min at room temperature to allow the interaction to take place between the protein and analytes of interest, followed by filtration using a 0.45 μm Polyvinylidene fluoride (PVDF) (Sigma-Aldrich, St. Louis, MO, USA). The concentrations of the pharmaceutical drugs were determined in terms of absorbance measurements using HPLC-DAD. The effects of the major experimental parameters (protein dosage, analyte concentration, and pH) on the removal of pharmaceutical drugs by proteins were investigated. Each sample was measured five times.

2.7. Data Analysis

The percentage removal of each of the pharmaceutical drugs was calculated based on the difference between the initial analyte concentration (C_0) (before removal) and the final analyte concentration (C_f) (after removal), which was obtained from the calibration curve for each of the analytes in the sample solutions, using the following formula:

$$\% \, removal = \frac{C_0 - C_f}{C_0} \times 100 \tag{1}$$

2.8. Sample Collection

Wastewater samples (effluent and influent) were collected from Daspoort, a WWTP located in Pretoria, one of the major cities in Gauteng Province, South Africa. The Daspoort WWTP discharges into a neighboring Apies River in its surroundings. The grab sampling method was used and the samples were collected in 2.5 L amber glass bottles, which had previously been washed and rinsed with ultra high pure UHP water and then flushed at least thrice with wastewater before collection. Each sample (influent and effluent) was collected in duplicate. All water samples were transported to the laboratory in cooler boxes packed with ice. Upon arrival, the samples were filtered through a Whatman (120 mm) filter paper using vacuum filtration and extracted immediately to avoid degradation. Thesamples were then stored at <4 °C in the dark and analyzed within 24 h.

3. Results and Discussion

3.1. Characterisation of Moringa Seed Protein Powder

The characterization of protein powder assists in gaining an understanding of features such as composition, structure, and various properties like physical and chemical properties. Moringa seed protein powder was characterized using FTIR to identify the functional groups of the active sites.

3.2. FTIR Characterisation of Moringa Protein before and after Removal of Antibiotics

As indicated in the FTIR spectra given in Figure 2, protein powder contains amine and amide functional groups at wave numbers 1647 cm^{-1}, 1541cm^{-1}, 1515 cm^{-1},and 1412 cm^{-1}associated with C=O stretch amide I, NH amide II, NH amide I bend, and C-N stretch amide III [33], respectively, which are active and able to bind with antibiotics. After the removal of the antibiotics, the interactions between the compound and the protein powder were indicated by the appearance of new peaks or the disappearance of peaks that were previously observed and a significant shift in peaks, as shown in Figure 2. The peaks at wave numbers 1541 $cm^{-1}$and 1529 cm^{-1}associated with NH amide II and NH amide I bending vibration, respectively, disappeared after loading antibiotics, while new peaks appeared at wave numbers 1565 and 3750 cm^{-1}. The band position shift from wave number 3207 to 3270 $cm^{-1}$was also observed.

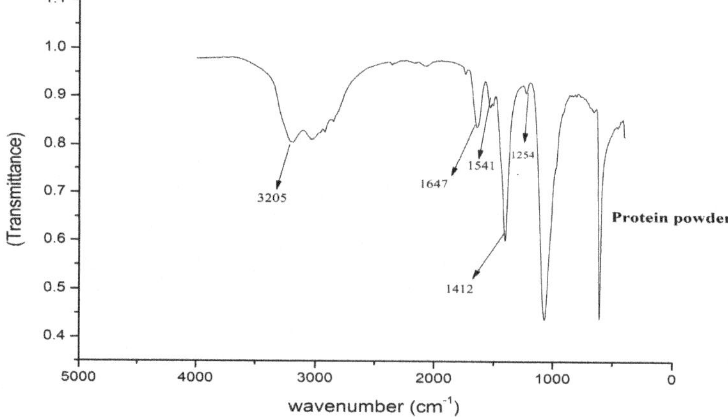

Figure 2. *Cont.*

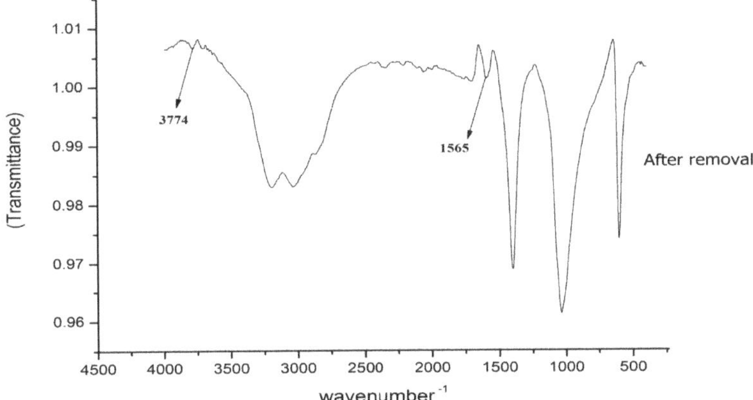

Figure 2. FTIR spectra of the water-soluble protein powder before and after the removal of multiclass antibiotics.

3.3. Removal of Antibiotics Using Moringa Protein

The water-soluble proteins extracted from *Moringa stenopetala* seed were used for the removal of selected antibiotics. Due to their complicated structure, proteins present in the seeds have a large number of functional groups; however, the amides and amines are the dominant functional groups. These functional groups were responsible for and played the major role in the removal of antibiotics. At different pH ranges, these functional groups behave differently. At pH values of between 3.5 and 10, proteins exist mainly in the zwitterionic form [33] while antibiotics are in the neutral and zwitterionic forms. Maximum removal was thus expected to occur when proteins are in the zwitterionic form and analytes in the neutral form, attributable to hydrogen bonding or electrostatic force of attraction between proteins and antibiotics. Different parameters that affect the efficiency of water-soluble proteins to remove antibiotics, such as protein dose, initial analyte concentration, and pH, were studied and the optimum conditions for each parameter were selected. For every analysis, each analyte was measured five times.

3.4. Effect of Initial Analyte Concentration

The analyte concentration in aqueous solution was one of the main factors that were found to affect the removal of antibiotics as shown in Figure 3. The concentration of antibiotics in the solution was varied from 0.1–1.5 mg L^{-1} while keeping the other parameters constant (i.e., protein dosage of 40 mg, pH at 5.5, volume 25 mL, contact time 30 min, and temperature at 23 °C). The experiment was run from the highest to the lowest concentration of antibiotics in order to determine the concentration where maximum removal of all analytes was achieved. The percentage removal for ten selected antibiotic drugs increased as the initial concentration decreased. This may have been due to the fact that at lower concentrations, there are sufficient active sites on the protein molecule for the analytes to occupy. However, the number of active sites is limited, and at higher concentrations, the active sites on proteins are rapidly occupied, and hence ions of selected analytes are left unbound in solution as a result of the saturation of binding sites. The maximum percentage removal obtained was 89.3 $\pm$ 0.05, 94.2 $\pm$ 0.02, 95.0 $\pm$ 0.02, 92.9 $\pm$ 0.05, 83.9 $\pm$ 0.06, 84.6 $\pm$ 0.04, 85.5 $\pm$ 0.06, 96.7 $\pm$ 0.01, 92.7 $\pm$ 0.07, and 84.9 $\pm$ 0.02 for sulphanilamide, marbofloxacin, ciprofloxacin, danofloxacin, oxytetracycline, sulphamerazine, sulphamonomethoxine, sulphamethoxazole, tylosin, and sulphadimethoxine, respectively.

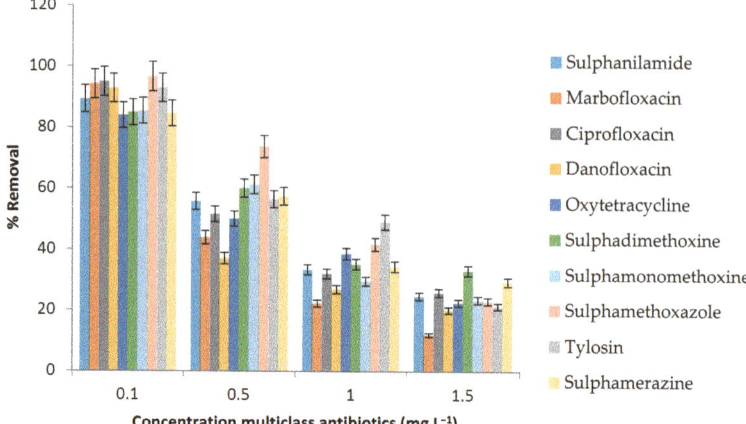

Figure 3. Effect of initial concentration on percentage removal of antibiotics using water-solubleprotein powder (n = 5).

3.5. Effect of Adsorbent Dosage

The amount of protein dosage was found to be an important parameter that affected the removal process, as shown in Figure 4. The removal was investigated by varying the dosage from 10 mg to 50 mg while keeping the other parameters constant (pH 5.5, analyte concentration of 0.1 mg L^{-1}, volume of 25 mL, contact time 30 min, and temperature at 25 °C). As the protein dosage was increased, the efficiency to remove antibiotics significantly increased, due to the increase in the number of available active sites responsible for the removal of antibiotics. The removal was also affected by the chemical structure and size of antibiotics. Maximum percentages removal of 88.8 ± 0.08, 94.2 ± 0.03, 94.2 ± 0.01, 91.9 ± 0.02, 89.7 ± 0.05, 87.6 ± 0.09, 85.2 ± 0.02, 96.3 ± 0.02, 91.5 ± 0.02, and 85.5 ± 0.02% were achieved for sulphanilamide, marbofloxacin, ciprofloxacin, danofloxacin, oxytetracycline, sulphamerazine, sulphamonomethoxine, sulphamethoxazole, tylosin, and sulphadimethoxine, respectively, at a protein dosage of 40 mg.

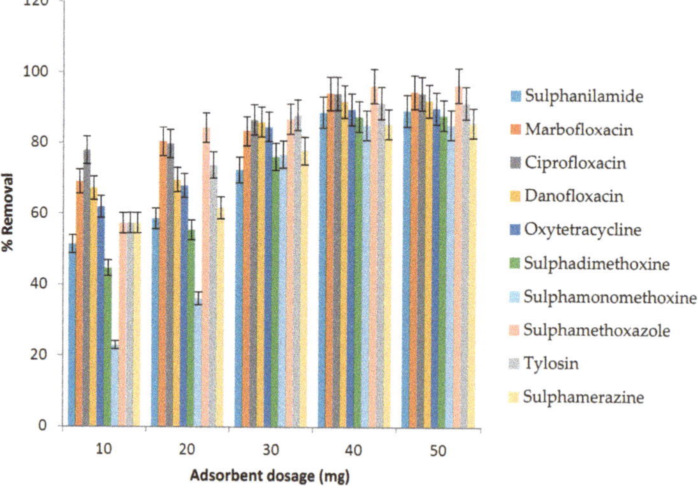

Figure 4. Effect of adsorbent dosage on percentage removal of antibiotics using water-soluble protein powder (n = 5).

3.6. Effect of pH

The percentage removal of selected antibiotics using water-soluble protein as a function of pH was studied in acidic, neutral, and basic media, as shown in Figure 5, while keeping other parameters constant (protein dosage 40 mg, analyte concentration of 0.1 mg L^{-1}, volume of 25 mL, contact time 30 min, and temperature at 25 °C. From the results, it was observed that the removal of antibiotics was strongly pH dependent. The pH does not only affect the property of analytes to be removed, but also affects the removal efficiency of the material as the shape and charge properties of active sites are affected by pH changes. The removal of ten selected antibiotics (sulphanilamide, marbofloxacin, ciprofloxacin, danofloxacin, oxytetracycline, sulphamerazine, sulphacetamide, sulphamonomethoxine, sulphamethoxazole, tylosin, and sulphadimethoxine) was significantly affected as the pH changed. The amine groups of fluoroquinolones (marbofloxacin, ciprofloxacin, and danofloxacin) at pH values below their pK_{a1} value (≤2.5) were protonated and positive charges were predominated. Protein also shows similar properties at pH values below their pK_{a1} value (≤2.5) [33], which may cause electrostatic repulsion between the protein and fluoroquinolones, resulting in a lower removal of fluoroquinolones. As the pH was increased to a range betweenpK_{a1} and pK_{a2}, both fluoroquinolones and proteins existed as zwitterions and maximum removal was observed due to hydrogen and electrostatic interaction. Further increases in pH to above 9 created a dominance of negative ions in both fluoroquinolones and protein molecules which might cause an electrostatic repulsion and would result in the poor removal of fluoroquinolones. The removal of sulphonamides (sulphanilamide, sulphadimethoxine, sulphacetamide, sulphamonomethoxine, sulphamethoxazole, and sulphamerazine) using water-soluble protein also proceeded in a similar way as that of fluoroquinolones due to the protonation of the amide group of sulphonamides at pH values below their pK_{a1} value (≤2.5). On the other hand, at pH values of between 2.5 and 6, sulphonamides mainly exist as neutral compounds. Maximum removal was observed for all sulphonamides when proteins were in the zwitterionic form and sulphonamides were in the neutral form. These findings are in agreement with the results reported by Yang et al. [32]; according to their studies, the maximum removal of sulphonamides using activated sludge was observed when sulphonamides were in the neutral form. Further increases in pH resulted in a decrease in the removal of sulphonamides, because at pH values above their pK_{a2}value, sulphonamides as well as proteins were negatively charged, which would create a stronger repulsion between the analyte and the adsorbent.

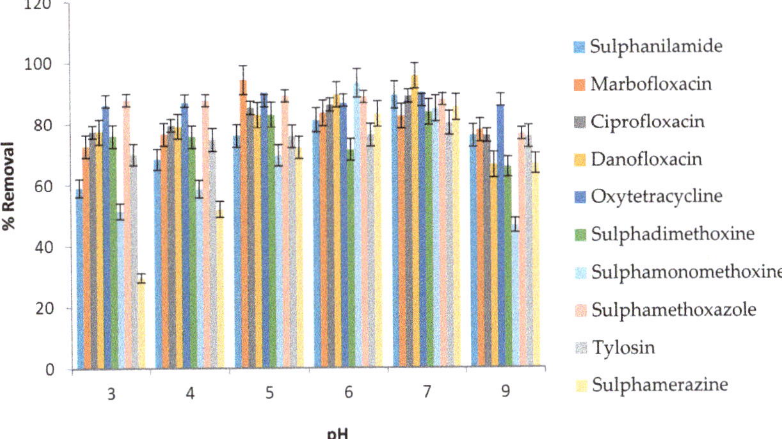

Figure 5. Effect of pH on the percentage removal of antibiotics using water-soluble protein powder (n = 5).

Oxytetracycline showed the same trend as sulphonamides and fluoroquinolones; however, it has three pK_a values. It exists as $X^0Y^+Z^0$ at a low pK_a value (3.3) and dissociates into $X^-Y^+Z^0$. At pK_{a2} its dissociation is from $X^-Y^+Z^0$ to either $X^-Y^+Z^-$or $X^-Y^0Z^0$, and then it dissociates into$X^-Y^0Z^-$ [36]. At a pH value below the pK_{a1}, quaternary amine was protonated, and cationic species became dominant. At pH values between pK_{a1} and pK_{a2}, the solution was mainly governed by neutral or zwitterionic species because of the protonation of tricarbonyl amide and deprotonation of phenolic diketone. With pH values higher than 10, anionic charges were predominant in oxytetracycline due to the deprotonation of all the functional groups. The maximum removal was observed in the neutral or zwitterionic state as a result ofhydrogen bonding and electrostatic interaction.

Tylosin tartrate which is group of macrilide, has two pK_a values (pK_{a1} = 3.3 and pK_{a2} = 7.7) and its protonation, deprotonation, and removal efficiency were the same as those of the other groups. At a pH value below pK_{a1}, the tartaric acid and tertiary amine were protonated, and as a result, the positive ions predominated; however, negative ions predominated at pH values above pK_{a2}.At a pH value below pK_{a2} and above pK_{a1}, zwitterions predominated and that is where the maximum removal was observed.

3.7. *Application of Method on Real Wastewater Sample*

The optimized method (protein dosage 40 mg, pH 6, analyte concentration of 0.1 mg L^{-1}, volume of 25 mL, contact time 30 min, and temperature at 25 °C) developed to evaluate water-soluble protein for the removal of selected antibiotics from a standard mixture solution was applied to wastewater after spiking 25 mL of the real wastewater sample with 2 mg L^{-1} standard solutions in order to obtain an effective concentration of 0.1 mg L^{-1}. The effluent and influent samples were spiked as the analytes of interest could not be detected using HPLC-DAD without preconcentration. As shown in Table 2, the results revealed that the developed method was found applicable for the removal of antibiotic drugs from the real wastewater sample. The maximum percentage removal obtained was in the range of 72.4 ± 0.32–92.5 ± 0.84 and 70.4 ± 0.82–91.5 ± 0.71 for effluent and influent, respectively. A slight decrease in the removal of the antibiotics in the real wastewater sample was observed compared to the removal obtained when using a standard mixture solution. The decrease in removal could be due to the presence of different competing ions in the real wastewater sample, which would compete for the available active sites on the protein. The results, however, confirmed the removal of antibiotic drugs from wastewater using water-soluble proteins extracted from *Moringa* seeds.

Table 2. Removal efficiency of water-soluble proteins in a real wastewater sample spiked with 2 mg L^{-1} of the standard solution (effective concentration was 0.1 mg L^{-1}) (n = 5).

Analytes	Retention Time	Concentration (mg L^{-1}) before Removal	Percentage Removal in the Ultrahigh Purity Water (%)	Percentage Removal in Effluent (%)	Percentage Removal of Influent (%)
Sulphanilamide	1.62	0.1	88.8 ± 0.05	72.4 ± 0.32	71.3 ± 0.56
Marbofloxacin	2.99	0.1	94.2 ± 0.05	88.2 ± 0.45	85.2 ± 0.66
Ciprofloxacin	3.35	0.1	94.2 ± 0.02	88.3 ± 0.56	83.5 ± 0.45
Danofloxacin	3.60	0.1	95.4 ± 0.12	89.5 ± 0.44	87.2 ± 0.85
Oxytetracycline	4.09	0.1	89.7 ± 0.04	92.5 ± 0.84	91.5 ± 0.71
Sulphamerazine	4.32	0.1	87.6 ± 0.05	74.4 ± 0.52	76.2 ± 0.32
Sulphamonomethoxine	5.69	0.1	85.2 ± 0.01	74.2 ± 0.32	70.1 ± 0.51
Sulfamethoxazole	6.28	0.1	96.3 ± 0.03	89.0 ± 0.56	83.7 ± 0.61
Tylosin tartrate	6.99	0.1	91.5 ± 0.01	86.8 ± 0.84	81.9 ± 0.55
Sulphadimethoxine	7.64	0.1	85.5 ± 0.01	74.7 ± 0.56	70.4 ± 0.82

4. Conclusions

In this study, the water-soluble protein powder extracted from the *Moringa stenopetala* seeds was characterized and the removal behavior was studied for ten selected antibiotic drugs (sulphanilamide, marbofloxacin, ciprofloxacin, danofloxacin, oxytetracycline, sulphamerazine, sulphamonomethoxine, sulphamethoxazole, tylosin tartrate, and sulphadimethoxine) using a known concentration of standard solution mixture prepared in the laboratory.

Different parameters that affected the removal efficiency of water-soluble proteins, such as protein dosage, initial analyte concentration, and pH, were studied and the optimum conditions for each parameter were selected. The optimum conditions for the removal of ten selected antibiotics usingwater-soluble proteins were a protein dosage of 40 mg, an initial antibiotic concentration of 0.1 mg L^{-1}, pH at 7, contact time 30 min, and volume 25 mL. The developed and optimized method was applied on wastewater. The simultaneous removal of selected antibiotics was investigated, and the results obtained confirmed that the water-soluble proteins extracted from *Moringa stenopetala* seeds are potentially useful to remove antibiotics from synthetic wastewater and real wastewater. The maximum percentage removal obtained using the developed method was in the range of 85.2–96.3% for sulphanilamide, marbofloxacin, ciprofloxacin, danofloxacin, oxytetracycline, sulphadimethoxine, sulphamonomethoxine, sulphamethoxazole, and tylosin tartrate. The percentage removal in the range of 72.4–92.5% and 70.4–91.5% was observed when the developed method was applied to the real wastewater sample (effluent and influent) collected from the wastewater treatment plant. Therefore, the developed method for the removal of selected antibiotics using water-soluble proteins from *Moringa stenopetala* seeds was simple, cost-effective, environmentally friendly, and easily applicable to monitor environmental pollution. Thus, *Moringa* can be regarded as a multipurpose tree (for medicinal purposes, food, and wastewater treatment), and as the results of this study indicate, its products can assist in keeping the environment free from pollution. Cultivating and managing these useful trees could also make a contribution towards reducing deforestation.

Author Contributions: Conceptualization, Investigation, Methodology and Writing original draft, T.G.K; Editing and Supervising, S.D.; and Editing and Supervising, M.M.N.

Funding: This research received no external funding.

Acknowledgments: The authors would like to acknowledge with thanks the contribution made by the Chemistry Department of the University of South Africa (UNISA) in making the equipment available and providing all the necessary chemicals.

Conflicts of Interest: The authors declare no conflict of interest.

References

1. Kümmerer, K. Antibiotics in the aquatic environment—A review—Part I. *Chemosphere* **2009**, *75*, 417–434. [CrossRef] [PubMed]
2. Thiele-Bruhn, S. Pharmaceutical antibiotic compounds in soils—A review. *J. Plant Nutr. Soil Sci.* **2003**, *166*, 145–167. [CrossRef]
3. Glassmeyer, S.T.; Hinchey, E.K.; Boehme, S.E.; Daughton, C.G.; Ruhoy, I.S.; Conerly, O.; Daniels, R.L.; Lauer, L.; McCarthy, M.; Nettesheim, T.G.; et al. Disposal practices for unwanted residential medications in the United States. *Environ. Int.* **2009**, *35*, 566–572. [CrossRef] [PubMed]
4. Yan, C.; Yang, Y.; Zhou, J.; Liu, M.; Nie, M.; Shi, H.; Gu, L. Antibiotics in the surface water of the Yangtze Estuary: Occurrence, distribution and risk assessment. *Environ. Pollut.* **2013**, *175*, 22–29. [CrossRef] [PubMed]
5. Lee, B. Removal of Antibiotics from Contaminated Waters Using Natural Zeolite. Master's Thesis, The City University of New York, New York, NY, USA, August 2012.
6. Karthikeyan, K.G.; Meyer, M.T. Occurrence of antibiotics in wastewater treatment facilities in Wisconsin, USA. *Sci. Total Environ.* **2006**, *361*, 196–207. [CrossRef] [PubMed]
7. Watkinson, A.J.; Murby, E.J.; Kolpin, D.W.; Costanzo, S.D. The occurrence of antibiotics in an urban watershed: From wastewater to drinking water. *Sci. Total Environ.* **2009**, *407*, 2711–2723. [CrossRef] [PubMed]
8. Xu, W.; Zhang, G.; Zou, S.; Li, X.; Liu, Y. Determination of selected antibiotics in the Victoria Harbour and the Pearl River, South China using high-performance liquid chromatography-electrospray ionization tandem mass spectrometry. *Environ. Pollut.* **2007**, *145*, 672–679. [CrossRef] [PubMed]

9. Chang, X.; Meyer, M.T.; Liu, X.; Zhao, Q.; Chen, H.; Chen, J.A.; Qiu, Z.; Yang, L.; Cao, J.; Shu, W. Determination of antibiotics in sewage from hospitals, nursery and slaughter house, wastewater treatment plant and source water in Chongqing region of Three Gorge Reservoir in China. *Environ. Pollut.* **2010**, *158*, 1444–1450. [CrossRef] [PubMed]
10. Berglund, B.; Gengler, S.; Batoko, H.; Wattiau, P.; Errampalli, D.; Leung, K.; Cassidy, M.B.; Kostrzynska, M.; Blears, M.; Lee, H.; et al. Environmental dissemination of antibiotic resistance genes and correlation to anthropogenic contamination with antibiotics. *J. Microbiol. Methods* **2015**, *113*, 28564. [CrossRef]
11. Nghiem, L.D.; Schäfer, A.I.; Elimelech, M. Removal of natural hormones by nanofiltration membranes: Measurements, modeling, and mechanisms. *Environ. Sci. Technol.* **2004**, *38*, 1888–1896. [CrossRef]
12. Milić, N.; Milanović, M.; Letić, N.G.; Sekulić, M.T.; Radonić, J.; Mihajlović, I.; Miloradov, M.V. Occurrence of antibiotics as emerging contaminant substances in aquatic environment. *Int. J. Environ. Health Res.* **2013**, *23*, 296–310. [CrossRef] [PubMed]
13. Xu, W.; Zhang, G.; Li, X.; Zou, S.; Li, P.; Hu, Z.; Li, J. Occurrence and elimination of antibiotics at four sewage treatment plants in the Pearl River Delta (PRD), South China. *Water Res.* **2007**, *41*, 4526–4534. [CrossRef] [PubMed]
14. Kemper, N. Veterinary antibiotics in the aquatic and terrestrial environment. *Ecol. Indic.* **2008**, *8*, 1–13. [CrossRef]
15. Terzić, S.; Senta, I.; Ahel, M.; Gros, M.; Petrović, M.; Barcelo, D.; Müller, J.; Knepper, T.; Martí, I.; Ventura, F.; et al. Occurrence and fate of emerging wastewater contaminants in Western Balkan Region. *Sci. Total Environ.* **2008**, *399*, 66–77. [CrossRef]
16. Watkinson, A.J.; Murby, E.J.; Costanzo, S.D. Removal of antibiotics in conventional and advanced wastewater treatment: Implications for environmental discharge and wastewater recycling. *Water Res.* **2007**, *41*, 4164–4176. [CrossRef]
17. Le-Minh, N.; Khan, S.J.; Drewes, J.E.; Stuetz, R.M. Fate of antibiotics during municipal water recycling treatment processes. *Water Res.* **2010**, *44*, 4295–4323. [CrossRef]
18. Elmolla, E.S.; Chaudhuri, M. Photocatalytic degradation of amoxicillin, ampicillin and cloxacillin antibiotics in aqueous solution using UV/TiO_2 and UV/H_2O_2/TiO_2 photocatalysis. *Desalination* **2010**, *252*, 46–52. [CrossRef]
19. González, O.; Sans, C.; Esplugas, S. Sulfamethoxazole abatement by photo-Fenton. Toxicity, inhibition and biodegradability assessment of intermediates. *J. Hazard. Mater.* **2007**, *146*, 459–464. [CrossRef]
20. Avisar, D.; Primor, O.; Gozlan, I.; Mamane, H. Sorption of sulfonamides and tetracyclines to montmorillonite clay. *Water. Air. Soil Pollut.* **2010**, *209*, 439–450. [CrossRef]
21. Wang, Z.; Pan, B.; Xing, B. Norfloxacin sorption and its thermodynamics on surface modified CNTs. *Environ. Sci. Technol.* **2010**, *44*, 978–984. [CrossRef]
22. Kim, Y.; Bae, J.; Park, J.; Suh, J.; Lee, S.; Park, H.; Choi, H. Removal of 12 selected pharmaceuticals by granular mesoporous silica SBA-15 in aqueous phase. *Chem. Eng. J.* **2014**, *256*, 475–485. [CrossRef]
23. Liu, Z.; Zhou, X.; Chen, X.; Dai, C.; Zhang, J.; Zhang, Y. Biosorption of clofibric acid and carbamazepine in aqueous solution by agricultural waste rice straw. *J. Environ. Sci.* **2013**, *25*, 2384–2395. [CrossRef]
24. Kebede, T.G.; Mengistie, A.A.; Dube, S.; Nkambule, T.T.I.; Nindi, M.M. Study on adsorption of some common metal ions present in industrial effluents by Moringa stenopetala seed powder. *J. Environ. Chem. Eng.* **2018**, *6*, 1378–1389. [CrossRef]
25. Kebede, T.G.; Dube, S.; Nindi, M.M. Fabrication and characterization of electrospun nanofibers from moringa stenopetala seed protein. *Mater. Res. Express* **2018**, *5*, 125015. [CrossRef]
26. Schwarz, J.; Thiele-Bruhn, S.; Eckhardt, K.-U.; Schulten, H.-R. Sorption of Sulfonamide Antibiotics to Soil Organic Sorbents: Batch Experiments with Model Compounds and Computational Chemistry. *ISRN Soil Sci.* **2012**, *2012*, 1–10. [CrossRef]
27. Bidgood, T.L.; Papich, M.G. Plasma and interstitial fluid pharmacokinetics of enrofloxacin, its metabolite ciprofloxacin, and marbofloxacin after oral administration and a constant rate intravenous infusion in dogs. *J. Vet. Pharmacol. Ther.* **2005**, *28*, 329–341. [CrossRef]
28. Şanli, N.; Şanli, S.; Özkan, G.; Denizli, A. Determination of pKa values of some sulfonamides by LC and LC-PDA methods in acetonitrile-water binary mixtures. *J. Braz. Chem. Soc.* **2010**, *21*, 1952–1960. [CrossRef]
29. Leo, A.; Hansch, C.; Elkins, D. Partition coefficients and their Uses. *Chem. Rev.* **1971**, *71*, 525. [CrossRef]

30. Heibrt, B.J.; Dorsey, J. mOctanol-Water Partition Coefficient Estimation by Micellar Electrokinetic Capillary Chromatography. *Anal. Chem.* **1995**, *67*, 744–749.
31. Qiang, Z.; Adams, C. Potentiometric determination of acid dissociation constants (pK a) for human and veterinary antibiotics. *Water Res.* **2004**, *38*, 2874–2890. [CrossRef]
32. Yang, S.F.; Lin, C.F.; Lin, A.Y.; Hong, P.K.A. Sorption and biodegradation of sulfonamide antibiotics by activated sludge: Experimental assessment using batch data obtained under aerobic conditions. *Water Res.* **2011**, *45*, 3389–3397. [CrossRef] [PubMed]
33. Kebede, T.G.; Dube, S.; Nindi, M.M. Removal of non-steroidal anti-in fl ammatory drugs (NSAIDs) and carbamazepine from wastewater using water-soluble protein extracted from Moringa stenopetala seeds. *J. Environ. Chem. Eng.* **2018**, *6*, 3095–3103. [CrossRef]
34. Ndabigengesere, A.; Narasiah, K.S.; Talbot, B.G. Active agents and mechanism of coagulation of turbid waters using Moringa oleifera. *Water Res.* **1995**, *29*, 703–710. [CrossRef]
35. Ndabigengesere, A.; Narasiah, K.S. Quality of water treated by coagulation using Moringa oleifera seeds. *Water Res.* **1998**, *32*, 781–791. [CrossRef]
36. Choi, K.J.; Son, H.J.; Kim, S.H. Ionic treatment for removal of sulfonamide and tetracycline classes of antibiotic. *Sci. Total Environ.* **2007**, *387*, 247–256. [CrossRef] [PubMed]

Article

Acetaminophen Removal from Water by Microalgae and Effluent Toxicity Assessment by the Zebrafish Embryo Bioassay

Carla Escapa [1,2], Ricardo N. Coimbra [3], Teresa Neuparth [2], Tiago Torres [2], Miguel M. Santos [2,4,*] and Marta Otero [3,5,*]

1. IMARENABIO-Institute of Environment, Natural Resources and Biodiversity, Department of Applied Chemistry and Physics, Universidad de León, Av. Portugal s/n, 24071 León, Spain; carla.escapa@unileon.es
2. CIMAR/CIIMAR-Interdisciplinary Centre of Marine and Environmental Research, Endocrine Disruptors and Emerging Contaminants Group, University of Porto, Av. General Norton de Matos s/n, 4450-208 Matosinhos, Portugal; tneuparth@ciimar.up.pt (T.N.); torres_tapt@hotmail.com (T.T.)
3. Department of Environment and Planning, University of Aveiro, Campus Universitário de Santiago, 3810-193 Aveiro, Portugal; ricardo.coimbra@ua.pt
4. FCUP-Department of Biology, Faculty of Sciences, University of Porto, Rua do Campo Alegre, 4169-007 Porto, Portugal
5. CESAM-Centre for Environmental and Marine Studies, Campus Universitário de Santiago, 3810-193 Aveiro, Portugal

* Correspondence: santos@ciimar.up.pt (M.M.S.); marta.otero@ua.pt (M.O.)

Received: 8 August 2019; Accepted: 12 September 2019; Published: 15 September 2019

Abstract: In this work, zebrafish embryo bioassays were performed to assess the efficiency of microalgae in the removal of acetaminophen from water. *Chlorella sorokiniana* (CS), *Chlorella vulgaris* (CV) and *Scenedesmus obliquus* (SO) were the strains used for water treatment. Toxic effects on zebrafish embryo caused by effluents from microalgae treatment were compared with those observed under exposure to experimental solutions with known concentrations of acetaminophen. The three microalgae strains allowed for the reduction of acetaminophen concentration and its toxic effects, but CS was the most efficient one. At the end of the batch culture, a 67% removal was provided by CS with a reduction of 62% in the total abnormalities on the exposed zebrafish embryo. On the other hand, toxic effects observed under exposure to effluents treated by microalgae were alike to those determined for acetaminophen experimental solutions with equivalent concentration. Thus, it may be inferred that microalgae biodegradation of acetaminophen did not involve an increased toxicity for zebrafish embryo.

Keywords: phyco-remediation; algae; wastewater; emerging contaminants; paracetamol; *Danio rerio*

1. Introduction

Pharmaceuticals belong to the class of emerging contaminants and, over the last decades, their presence and persistence in the environment has caused great concern due to the threat that they represent for aquatic and terrestrial life [1]. Wastewater is a main source of pharmaceuticals in the aquatic environment, a main difficulty in wastewater treatment being that these pollutants are single compounds with an individual behaviour and represent only a minor part of the wastewater organic load [2]. These singularities have called to new approaches in wastewater treatment to limit the discharge of pharmaceuticals in receiving waters [2].

On the other hand, the application of microalgae-based water treatments is raising scientific interest. Among others, advantages of microalgae treatment include photoautotrophic growth, relatively small amounts of operational inputs, eco-friendliness, CO_2 sequestration, high-value by-products such as nutraceuticals and cosmetics and simultaneous production of low-value foodstuff for aquaculture. Moreover, algal biomass resulting from water treatment may be also used as biosorbent, fertilizer, animal feed, for energy valorisation or biofuels production [3–6]. Thus, integrating microalgae culture and wastewater treatment is a sustainable alternative that comprises energy and resources recovery [7].

Microalgae cultivation systems can be classified as open or closed, like photobioreactors (PBRs) [8]. The latter allow the guarantee of mono-cultures and a tighter control of operation conditions, their application for water treatment (with simultaneous CO_2 sequestration) having been initiated at the Carnegie Institute of Washington in 1953 [9]. Since then, the study of microalgae for water treatment has been mainly focused on the removal of nutrients [9–12], although these systems have been also used for the removal of heavy metals [9,11,12] and, more recently, pharmaceuticals [6,9,11,12].

Despite the scale-up challenges that microalgae water treatment still faces, high removal percentages have been attained for most of target pharmaceuticals at laboratory scale [11,12]. In fact, it is well-known that micro-organisms such as bacteria, fungi, protozoa and microalgae tend to degrade organic contaminants to a much larger degree than humans and animals [13]. Microalgae removal of pharmaceuticals occur by bioaccumulation, bioadsorption and, especially, by biodegradation [6]. However, pharmaceuticals degradation may result in the generation of transformation products (TPs) that can be equally or even more toxic than parent compounds [14], which casts doubt on the convenience of using microalgae-based treatments in the removal of pharmaceuticals. Monitoring TPs from microalgae water treatment, which would dispel any doubts about this issue, is impractical. In fact, due to analytic limitations, few authors have attempted to determine TPs from microalgae biodegradation of pharmaceuticals and hormones [15–17]. Indeed, from a practical point of view, it is not realistic neither necessary to identify every possible TP for a given micropollutant [18].

In the literature, the evaluation of microalgae efficiency in the removal of pharmaceuticals is commonly expressed in percentage terms. Though this approach is used to assess the reduction of conventional pollutants in conventional water treatment plants, due to the possible generation of TPs, it is not conclusive in the case of micropollutants [18]. Given the unviability of full monitoring of TPs from the degradation of this sort of pollutants, complementing analytical measurements with toxicological data may be a suitable strategy for the evaluation of microalgae-based treatments efficiency. Furthermore, with ever-increasing environmental awareness, proving the capacity of microalgae-based treatments to reduce not only concentration, but also the associated toxicity effects of pharmaceuticals, would serve to encourage their practical implementation. Therefore, in this context, this work aimed at verifying if pharmaceutical removal by microalgae co-occurs with a reduction of toxicity effects.

Chlorella vulgaris, *Chlorella sorokiniana* and *Scenedesmus obliquus*, which are among the most commonly employed microalgae strains in water treatment, were used and compared in this work regarding their efficiency in the removal of acetaminophen and its toxic effects. Acetaminophen (paracetamol, acetyl-para-aminophenol, N-acetyl-p-aminophenol) was selected as the target pharmaceutical, since it is a widely used over-the-counter analgesic and antipyretic drug. Due to its extensive consumption, this pharmaceutical is mostly ubiquitous in influents to sewage treatment plant (STPs) and appreciable concentrations have been found in STP effluents and surface waters [19]. Regarding the toxicity evaluation of effluents from microalgae treatment, zebrafish embryo bioassays were carried out in this work. Despite higher animals traditionally being considered models of excellence for the evaluation of drugs toxicity, zebrafish has recently been presented as a reliable vertebrate model to determine developmental toxicity and general toxicity of drugs [20]. Zebrafish (*Danio rerio*) embryos represent an attractive model for toxicity studies on pharmaceuticals [21] and hold several practical advantages, namely their small size, large robustness, short life-cycle, great number of offspring, simple cost-effective management and reproduction at laboratory-scale and translucent eggs that allow for stereomicroscope monitoring of embryo development [22,23].

2. Material and Methods

2.1. Zebrafish (Danio rerio) Embryo Toxicity Bioassays

2.1.1. Fertilization and Collection of Zebrafish Embryos

Zebrafish adults were kept in a 70-L aquaria filled with freshwater at 28 ± 1 °C and under a 14:10 h light:dark photoperiod, as described by Soares et al. [24]. *Ad libitum* feeding of these adults was done twice a day, the feedstuff consisting of Tetramin (Tetra, Melle, Germany) supplemented with *Artemia* spp.

For spawning, adult males and females (2:1) were placed in 30-L breeding tanks overnight under the same water conditions above described. Then, ovulation and fertilization of the eggs were stimulated by the beginning of light period [24]. Next, fertilized eggs were collected from the bottom of the tank and washed with water several times in order to remove detritus and avoid micro-organisms' proliferation during the subsequent bioassays [25].

2.1.2. Experimental Solutions

In order to validate the applicability of *Danio rerio* embryo bioassays for the targeted aim, these were first carried out on experimental solutions of acetaminophen with concentrations in the range of influent and effluent concentrations in microalgae treatments, as described in Section 2.2. For this purpose, acetaminophen ($C_8H_9NO_2$, ≥99%) acquired from Sigma-Aldrich (Madrid, Spain) was used to make experimental solutions with the following concentrations: 25, 250, 2500, 6250, 12,500 and 25,000 $\mu g\ L^{-1}$. These solutions were prepared by dilution of acetaminophen in the standard microalgae culture medium Mann and Myers [26] and freshwater (at 1:1 ratio). Furthermore, an experimental and a solvent control were also tested. For each acetaminophen concentration, experimental and solvent control, six replicates were carried out. All the solutions and controls were daily prepared in order to guarantee the pharmaceutical concentration and to avoid micro-organisms proliferation.

2.1.3. Zebrafish Embryo Bioassays

The static-water renewal toxicological zebrafish bioassays were carried out following the ecotoxicity test guidelines of the Organization for Economic Cooperation and Development (OECD), Test No. 236: Fish Embryo Acute Toxicity (FET) Test [27] and Ribeiro et al. [25]. After embryos observation using a magnifying glass, 10 fertilized eggs were selected and randomly allocated into 24-wells plates filled with 2 mL of freshly prepared acetaminophen solution, experimental control or solvent control. The 24-wells plates were then incubated at 26.5 °C during 144 h and under the same photoperiod conditions as the zebrafish adults. The medium was renewed daily in order to maintain dissolved oxygen and acetaminophen nominal concentrations constant during the bioassay and to remove fungi or other organisms that could proliferate in the well.

The effects of exposure were assessed at four distinct periods described by Kimmel et al. [28] as representative of important steps of embryo development (embryo pictures at these periods are depicted in Figure S1): gastrula period (75% epiboly stage), pharyngula period (prim15–16), larval stage (protruding-mouth) and juvenile, which were respectively observed at 8, 32, 80 and 144 h post fertilization (hpf). At each observation time, non-viable embryos were removed if present, mortality rate was assessed and morphological abnormalities were rated as abnormalities in eyes, head, tail or yolk-sac, developmental delay, abnormal cells, pericardial oedema, opaque chorion, excess or lack of pigmentation, lateral position, reduced mobility and involuntary movements. The total abnormalities rate was recorded as the percentage of embryos with at least one of the referred abnormalities in comparison to the control. Also, 75% epiboly stage at 8 hpf, hatching rate at 80 hpf and larval length at 144 hpf were evaluated in accordance with FET 236 [27] and Torres et al. [23]. Observations were performed using an inverted microscope (Nikon Eclipse 5100T, Nikon Corporation, Tokyo, Japan) equipped with a digital camera (Nikon D5-Fi2, Nikon Corporation, Tokyo, Japan) and a microscope camera controller (Nikon's Digital Sight DS-U3, Nikon Corporation, Tokyo, Japan).

In all sets of experiments, it was verified that the control groups did not show a mortality rate higher than 10%, which ensures that zebrafish embryo bioassays were adequately carried out.

2.1.4. Statistical Analysis

Statistical analysis was performed with SPPS Statistics software (version 21.0) (International Business Machines Corporation (IBM), New York, NY, USA). Data were first tested for variance normality and homogeneity using Kolmogorov–Smirnov and Levene tests. If these assumptions were met, differences between treatments were tested for significance by means of one-way factorial ANOVA followed by Newman–Keuls multiple comparison test to compare the control groups and each of the exposed groups. If the homogeneity and normality were not met, data were analysed by the non-parametric Kruskal–Wallis test, followed by a multiple comparison rank test (U Mann–Whitney test). Results for mortality rate, morphological and developmental abnormalities were expressed as the mean ± standard error (SE) and differences were considered significant for $p \leq 0.05$. Statistical analysis was done at each observation time defined at the previous section. Experimental and solvent controls were grouped when no significant differences between them were detected.

2.2. *Acetaminophen Removal from Water by Microalgae*

Microalgae used in this work for the removal of acetaminophen from water were *Chlorella sorokiniana* (CS), *Chlorella vulgaris* (CV) and *Scenedesmus obliquus* (SO). The selection of these strains was based on their well-established use for water treatment. The culture of these microalgae was carried out under identical conditions in bubbling column PBRs (diameter = 4 cm, height = 30 cm), which were operated under batch conditions and at constant temperature (25 ± 1 °C), irradiance (370 μE m^{-2} s^{-1}), photoperiod (12:12), aeration with CO_2 enriched air (0.3 v/v/min) and pH (7.5 ± 0.5), as described in a previous work [29].

For each CS, CV and SO, PBRs were run in triplicate and simultaneously in order to evaluate the removal efficiency of acetaminophen. In these experiments, initial concentration of acetaminophen in the culture medium was 25,000 μg L^{-1}. Such a relatively high feeding concentration was needed to guarantee the applicability of the methodologies employed in this work. Negative controls (25,000 μg L^{-1} acetaminophen in culture medium, with no microalgae) and positive controls (microalgae in culture medium, with no acetaminophen) were also simultaneously run. During operation, a 5-mL aliquot was daily taken from each PBR (experiment, negative or positive control) in order to determine the concentration of microalgae biomass (C_b) and that of acetaminophen so to respectively monitor microalgae growing and removal of the target pharmaceutical.

2.2.1. Analytic Methods and Instrumentation

All aliquots were analysed for C_b and acetaminophen concentration. C_b was determined by optical density at 680 nm (OD_{680}) using a UV-visible spectrophotometer (BECKMAN DU640). Equation (1), which was established from preliminary experiments, was used for the calculation of C_b:

$$OD_{680} = 5.1834C_b + 0.0128, R^2 = 0.9983 \quad (1)$$

Acetaminophen concentration was analysed in a Waters HPLC 600 equipped with a 2487 dual λ absorbance detector. A Phenomenex Gemini-NX C18 column (5 μm, 250 mm × 4.6 mm) was used for the separation and the wavelength of detection was 246 nm. Before analysis, and in order to obtain clear samples for chromatographic analysis, aliquots were twice centrifuged at 7500 rpm for 10 min (SIGMA 2-16P centrifuge). The mobile phase consisted of acetonitrile/ultrapure water (30:70, v/v), which was filtered through a Millipore membrane (pore size = 0.45 μm) and degasified (for 30 min) before use. HPLC quality acetonitrile (CH_3CN) was purchased from DBH Prolabo Chemicals and ultrapure water was produced by a Millipore System.

2.2.2. Evaluation of Effluents Toxicity by the Zebrafish Embryo Bioassay.

When microalgae concluded the exponential growth phase, 75 mL of effluent were taken from each PBR, homogenized and twice centrifuged (7500 rpm) during 10 min in order to obtain clear samples. Next, the resulting supernatant was diluted (1:1) with freshwater to be used in zebrafish embryo bioassays. Such a dilution was set as the most favourable on the basis of a preliminary study using different dilutions with freshwater (1:7, 1:3, 1:1, 1:0), which was carried out to determine an appropriate ratio that allowed for the observation of effects in zebrafish embryo but avoided biased effects caused by the culture medium. Still, for additional validation of the absence of biased effects, bioassays were also carried out using a 1:3 dilution with freshwater. Effluents from the microalgae treatment (at each 1:1 and 1:3 dilution ratios with freshwater) were tested for zebrafish embryo toxicity following the procedure already described in Section 2.1.

3. Results

3.1. *Acetaminophen Experimental Solutions*

Results on the mortality and total abnormalities of zebrafish embryos exposed to acetaminophen experimental solutions are shown in Figure 1 while the observed abnormalities are depicted in Table 1.

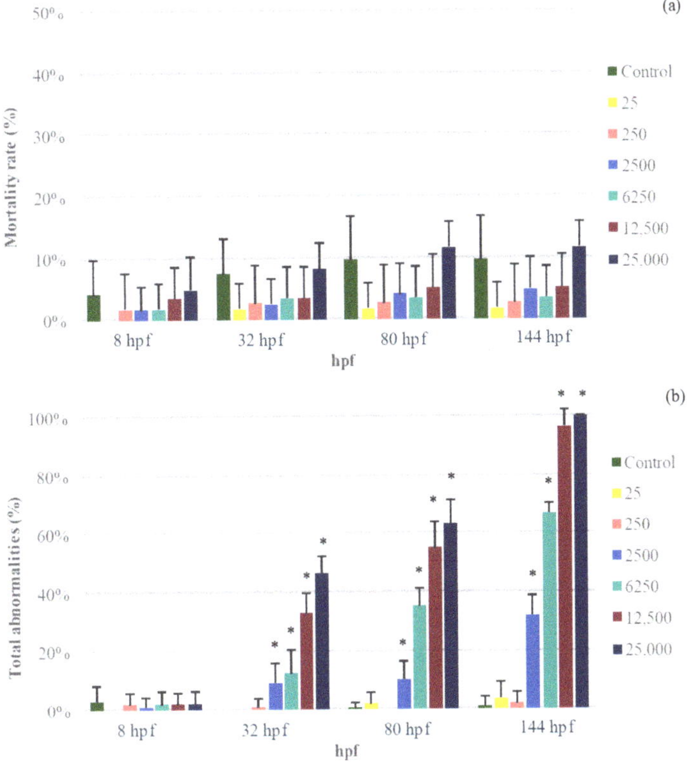

Figure 1. Mortality (**a**) and total abnormalities (**b**) observed for zebrafish embryos exposed to acetaminophen experimental solutions (25, 250, 2500, 6250, 12,500 and 25,000 μg L^{-1}) at the different observation times (8, 32, 80 and 144 h post fertilization (hpf)). Note: Mean results (n = 12 for control; n = 6 for exposed groups) are shown together with standard error (SE). Results significantly different from control ($p \leq 0.05$) are marked with a symbol (*).

Table 1. Effects of acetaminophen experimental solutions with different concentrations on zebrafish embryos at the observation time-points considered in this work.

Observation Time	Acetaminophen Concentration	75% Epiboly Rate	Developmental Delay	Lack of Pigmentation	Excess of Pigmentation	Lateral Position	Involuntary Movements	Larval Length (μm)
8 hpf	0 (control)	93.7 ± 7.6	3.2 ± 4.9					
	25	100 ± 0.0	0.0 ± 0.0					
	250	97.5 ± 6.2	0.8 ± 2.9					
	2500	96.8 ± 6.4	0.9 ± 3.2					
	6250	96.5 ± 5.5	1.9 ± 4.5					
	12,500	95.0 ± 5.5	0.0 ± 0.0					
	25,000	93.5 ± 8.1	0.0 ± 0.0					
32 hpf	0 (control)		0.0 ± 0.0	0.0 ± 0.0	0.0 ± 0.0			
	25		0.0 ± 0.0	0.0 ± 0.0	0.0 ± 0.0			
	250		0.0 ± 0.0	0.0 ± 0.0	0.0 ± 0.0			
	2500		0.0 ± 0.0	**9.2 ± 6.7**	0.0 ± 0.0			
	6250		0.0 ± 0.0	**12.4 ± 8.0**	0.0 ± 0.0			
	12,500		0.0 ± 0.0	**32.6 ± 6.8**	0.0 ± 0.0			
	25,000		0.0 ± 0.0	**46.5 ± 5.5**	0.0 ± 0.0			
80 hpf	0 (control)		0.0 ± 0.0	0.0 ± 0.0	0.0 ± 0.0			
	25		0.0 ± 0.0	0.0 ± 0.0	0.0 ± 0.0			
	250		0.0 ± 0.0	0.0 ± 0.0	0.0 ± 0.0			
	2500		0.0 ± 0.0	0.0 ± 0.0	9.4 ± 6.7			
	6250		0.0 ± 0.0	0.0 ± 0.0	**35.2 ± 5.8**			
	12,500		0.0 ± 0.0	0.0 ± 0.0	**54.6 ± 9.1**			
	25,000		0.0 ± 0.0	0.0 ± 0.0	**63.0 ± 8.4**			
144 hpf	0 (control)		0.0 ± 0.0	0.0 ± 0.0	0.0 ± 0.0	0.7 ± 2.8	0.0 ± 0.0	3861.23 ± 66.85
	25		0.0 ± 0.0	0.0 ± 0.0	0.0 ± 0.0	1.9 ± 4.5	0.0 ± 0.0	3874.83 ± 69.40
	250		0.0 ± 0.0	0.0 ± 0.0	0.0 ± 0.0	1.7 ± 3.9	0.0 ± 0.0	3900.42 ± 49.33
	2500		0.0 ± 0.0	0.0 ± 0.0	**31.9 ± 6.7**	0.9 ± 3.2	0.0 ± 0.0	3909.25 ± 68.96
	6250		0.0 ± 0.0	0.0 ± 0.0	**66.7 ± 3.7**	**22.8 ± 7.9**	0.0 ± 0.0	**3980.83 ± 89.77**
	12,500		0.0 ± 0.0	0.0 ± 0.0	**96.5 ± 5.5**	**64.8 ± 5.8**	**26.3 ± 5.5**	**4052.83 ± 44.20**
	25,000		0.0 ± 0.0	0.0 ± 0.0	**100 ± 0.0**	**61.2 ± 5.0**	**35.1 ± 7.6**	**4047.67 ± 78.22**

Note: Control and solvent control were grouped. In addition, the treatments with the same concentration from different sets of experiments (2500 and 250 μg L^{-1}) were grouped. Mean results (n = 12 for control; n = 6 acetaminophen exposed groups) are shown together with SE. Results significantly different from control ($p \leq 0.05$) are marked in bold.

As seen in Figure 1a, mean mortality rates were not larger than 12% in any case. Furthermore, the concentrations of acetaminophen here considered did not cause effects on the mortality at any observation time-point. Regarding total abnormalities, Figure 1b evidences that abnormalities were not observed until 32 hpf, their incidence significantly increasing at longer time-points. In any case, significance increases in total abnormalities were observed for acetaminophen initial concentrations higher than 250 μg L^{-1} at 32, 80 and 144 hpf.

Table 1 shows that the 75% epiboly rate and total abnormalities at 8 hpf did not show significant differences at any acetaminophen concentration in comparison with the control. At 32 hpf, lack of pigmentation was observed in the embryos exposed to concentrations equal or higher than 2500 μg L^{-1} ($p \leq 0.05$) in a range of 9.2% to 46.5%. However, no developmental delay was detected. After 48 h (80 hpf), abnormalities increased, and an excess of pigmentation was observed on larvae exposed to concentrations equal or higher than 2500 μg L^{-1} ($p \leq 0.05$), in a range of 10.2% to 63.0%. Still, the hatching rate was not affected by any of the acetaminophen tested concentrations.

At 144 hpf, the effects of the pharmaceutical on total abnormalities increased significantly in comparison with control ($p \leq 0.05$) (100% in the case of 25,000 μg L^{-1} acetaminophen). Moreover, some of the larvae remained in lateral position. Indeed, those larvae exposed to the highest concentrations of acetaminophen that remained in lateral position, also showed spasms or involuntary movements (>35% under 25,000 μg L^{-1}, >25% under 12,500 μg L^{-1}). Furthermore, the highest acetaminophen concentrations (≥6250 μg L^{-1}) caused a marginal increase in the larvae length ($p > 0.05$).

Overall, the above results point towards time and concentration dependence of acetaminophen effects on zebrafish embryo. Yet, effects on mortality rate were not registered under exposure to any of the considered acetaminophen concentrations.

3.2. *Acetaminophen Removal from Water by Microalgae*

For the strains here considered, results on the concentration of acetaminophen in PBRs and microalgae biomass throughout the batch culture are shown in Figure 2. Acetaminophen concentration was observed to decrease while microalgae biomass increased during the culture of the three strains used in this work. The steady state was reached after eight or nine days of cultivation. During the culture, the concentration of acetaminophen in the negative controls was stable. Therefore, it may be stated that, in the treatment experiments, the decrease of acetaminophen concentration was associated to the presence of microalgae.

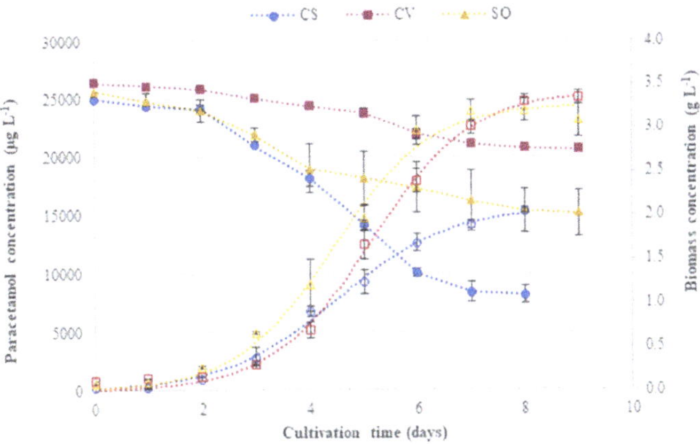

Figure 2. Concentration of acetaminophen (full symbols) and microalgae biomass (open symbols) in photobioreactors (PBRs) during the culture of *Chlorella sorokiniana* (CS), *Chlorella vulgaris* (CV) and *Scenedesmus obliquus* (SO). Note: Mean results ($n = 3$) are shown together with the corresponding SE.

As may be seen in Figure 2, the removal of acetaminophen by CS, with outlet concentrations of about 8200 µg L^{-1}, was more effective than that by SO or CV, which provided final acetaminophen concentrations around 15,200 and 20,700 µg L^{-1}, respectively. As for these concentrations in the effluents, the removal of acetaminophen at the end of the batch culture was 67% (CS), 39% (SO) and 17% (CV). Strain specific differences must be underneath these different efficiencies, which are not directly related with biomass growth, since the biomass of CS at the end of the batch (2 g L^{-1}) was lower than for SO and CV (about 3 g L^{-1}).

3.3. *Evaluation of Effluents Toxicity by the Zebrafish Embryo Bioassay*

Mortality and total abnormalities in zebrafish embryo exposed to treated effluents by CS, CV and SO are shown in Figure 3. Additionally, abnormalities and larval length registered at each observation time-point are specified in Table 2. Furthermore, the observed effects associated to exposure to effluents at 1:3 dilution are shown within the Supplementary Material (Table S1).

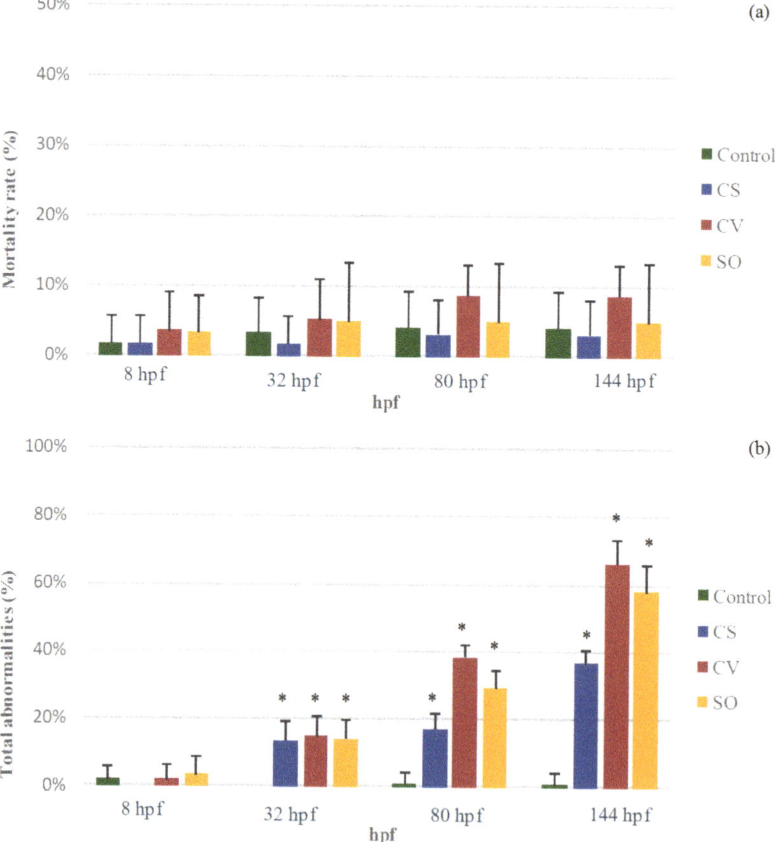

Figure 3. Mortality (**a**) and total abnormalities (**b**) observed for zebrafish embryos exposed to effluents from microalgae treatment for the removal of acetaminophen from water. Results are shown at the different observation times (8, 32, 80 and 144 hpf) for CS, CV and SO microalgae treatment. Note: Mean results (n = 12 for control; n = 6 for exposed groups) are shown together with SE. Results significantly different from control ($p \leq 0.05$) are marked with a symbol (*). In Figure 3a, the scale of the vertical axis was fitted (from 0% to 50%) for the visualization of results.

Table 2. Effects of exposure to microalgae treated effluents on zebrafish embryos exposed to treated effluents by CS, CV and SO embryos at the observation time-points considered in this work.

Observation Time	Effluent	75% Epiboly Rate	Developmental Delay	Lack of Pigmentation	Excess of Pigmentation	Lateral Position	Involuntary Movements	Larval Length (μm)
8 hpf	Control	96.6 ± 5.1	1.8 ± 4.1					
	CS	98.3 ± 4.1	0.0 ± 0.0					
	CV	94.6 ± 5.9	1.9 ± 4.5					
	SO	95.0 ± 5.5	3.3 ± 5.2					
32 hpf	Control		0.0 ± 0.0	0.0 ± 0.0	0.0 ± 0.0			
	CS		0.0 ± 0.0	**13.6 ± 5.3**	0.0 ± 0.0			
	CV		0.0 ± 0.0	**14.9 ± 5.8**	0.0 ± 0.0			
	SO		0.0 ± 0.0	**14.5 ± 5.2**	0.0 ± 0.0			
80 hpf	Control		0.0 ± 0.0	0.0 ± 0.0	0.0 ± 0.0			
	CS		0.0 ± 0.0	0.0 ± 0.0	**17.0 ± 4.6**			
	CV		0.0 ± 0.0	0.0 ± 0.0	**36.5 ± 2.7**			
	SO		0.0 ± 0.0	0.0 ± 0.0	**25.2 ± 8.4**			
144 hpf	Control		0.0 ± 0.0	0.0 ± 0.0	0.0 ± 0.0	0.0 ± 0.0	0.0 ± 0.0	3856.88 ± 45.84
	CS		0.0 ± 0.0	0.0 ± 0.0	**36.7 ± 3.7**	0.0 ± 0.0	0.0 ± 0.0	3862.85 ± 54.11
	CV		0.0 ± 0.0	0.0 ± 0.0	**64.6 ± 9.4**	**32.1 ± 6.2**	0.0 ± 0.0	**4067.03 ± 35.16**
	SO		0.0 ± 0.0	0.0 ± 0.0	**54.2 ± 8.1**	**22.9 ± 6.7**	0.0 ± 0.0	**4027.35 ± 42.54**

Note: Control and solvent control were grouped. Significant differences from control ($p \leq 0.05$) are marked in bold. Data are expressed as mean ± SE (n = 12 for control; n = 6 for effluents exposed groups).

As already observed for the acetaminophen experimental solutions (Figure 1a), no significant effects on mortality were either observed on zebrafish embryo after exposure to effluents from microalgae treatment (Figure 3a). With respect to total abnormalities incidence (Figure 3b), no differences with the control were observed for effluents at 8 hpf. However, at 32 hpf and longer time-points, exposure to effluents from microalgae treatment caused a significant increase of the percentage of total abnormalities in zebrafish embryo as compared with the control. Still, differences among the three microalgae strains were just revealed at 80 hpf and at 144 hpf, being higher at 144 hpf. In this regard, the average percentages of total abnormalities caused by effluents from CS, SO and CV at 144 hpf were 37%, 66% and 58%, respectively. These values are all comprised between the percentages determined for acetaminophen experimental solutions of 2500 and 6250 μg L^{-1} (Figure 1b) and evidence that microalgae treatment led to a reduction of total abnormalities incidence as compared with that observed for the initial concentration of acetaminophen (12,500 μg L^{-1} in Figure 1b, as for the 1:1 dilution). On the other hand, the percentages of total abnormalities caused by exposure to effluents (Figure 3b) were parallel with the efficiency of each microalgae strain in the removal of acetaminophen, which was referred in the previous section (67% (CS), 39% (SO) and 17% (CV)).

Concerning the anomalies caused in zebrafish embryo by the effluents from microalgae treatment, Table 2 shows that neither the 75% epiboly rate at 8 hpf nor the hatching rate at 80 hpf were altered in comparison to the control. Moreover, at 32 hpf a significant ($p \leq 0.05$) lack of pigmentation on embryos exposed to effluents was observed, with average incidence below 15% for the three microalgae treatments. At 80 hpf, the incidence of total abnormalities in embryos exposed to effluents increased (Figure 3b), although these abnormalities were exclusively excess of pigmentation (Table 2). Effluents from CV and SO showed higher mean of abnormalities (36% and 25%, respectively) than effluents from CS (17%). Still, it was at 144 hpf when effects of effluents on zebrafish embryos were more visible, which reflects the time dependence of effects that was already observed for acetaminophen experimental solutions.

As seen in Table 2, at 144 hpf, exposure to effluents from the three microalgae treatments caused an excess of pigmentation, which ranged from 37% (CS) to 65% (CV). In addition, embryo exposed to effluents from CV and SO remained in lateral position (mean incidence of 32% and 23%, respectively) and showed larger larval length. Therefore, the incidence of effects on zebrafish embryos has a parallelism with the final concentration of acetaminophen at the end of the culture, as shown in Figure 2 (CV > SO > CS). This is further corroborated by the results obtained under the 1:3 dilution here considered (Table S1), evidencing that the larger the acetaminophen concentration in the effluent, the larger the incidence of effects on zebrafish embryos.

4. Discussion

Acetaminophen experimental solutions tested in this work (25 to 25,000 μg L^{-1}) did not affect the mortality of exposed zebrafish embryo at any of the developmental stages considered. Coincidently with these results, no significant effects on mortality were noticed by Nogueira et al. [30] on zebrafish embryo subjected 5 to 3125 μg L^{-1} acetaminophen up to 96 hpf. Under a narrower exposure concentration (5 to 500 μg L^{-1} acetaminophen in 0.005% dimethyl sulphoxide (DMSO)) and with observation times until 96 hpf, Xia et al. [31] also did not observe an increase in the mortality rate of zebrafish embryo. Likewise, but at higher concentrations (151 to 756,000 μg L^{-1}), Pandya et al. [32] did not detect mortality effects in zebrafish embryos exposed to acetaminophen solutions in experiments with a duration of five days post-fertilization (dpf). Similarly, Peng et al. [33], using a transgenic zebrafish line Tg (wt1b: GFP) for the study of nephrotoxicity, observed no effects (12–72 hpf) on the survival of embryos exposed to even higher concentrations of acetaminophen (340, 3400 and 6804 mg L^{-1}). Contrarily, under lower concentrations (1, 5, 10, 50 and 100 μg L^{-1} acetaminophen dissolved in 0.1% aqueous ethanol), David and Pancharatna [34] observed significantly increased mortality rates for concentrations equal or higher than 5 μg L^{-1}.

In the present study, under acetaminophen exposure, embryo abnormalities were registered in zebrafish for concentrations equal or higher than 2500 μg L^{-1} at observation time-points equal or longer than 32 hpf. These abnormalities were lack of pigmentation at 32 hpf, excess of pigmentation at 80 hpf, and, at 144 hpf, excess of pigmentation, lateral position, involuntary movements and larger larval length. Furthermore, the incidence of abnormalities increased with observation time and concentration. Likewise, no effects were observed on zebrafish embryo by Nogueira et al. [30] after the first 24 h of exposure to 5 to 3125 μg L^{-1} acetaminophen. However, although the dose-response was not so clear as in this work, at 48 hpf and longer observation times, abnormalities were noticed for the embryos and larvae exposed to acetaminophen, namely lack of pigmentation (48 and 72 hpf), abnormal bending of the spine (72 and 96 hpf) and alteration of the larvae equilibrium (72 hpf) [30]. David and Pancharatna [34] also observed that the distribution of pigment in zebrafish embryo was dose-dependent, detecting a lack of pigmentation under 50 and 100 μg L^{-1} acetaminophen. Moreover, these authors [34] found that larvae exposed to 10, 50 and 100 μg L^{-1} acetaminophen showed altered swimming behaviour such as vibratory/shivering. Differently, Xia et al. [31] did not observe significant impacts on zebrafish embryo movement from acetaminophen exposure to 5, 50 and 500 μg L^{-1}. The latter study is in agreement with observations in this work, since involuntary movements occurred just for embryos exposed to the highest acetaminophen concentrations, namely 12,500 and 25,000 μg L^{-1}.

During microalgae cultivation in PBRs for water treatment, CS was clearly the most effective in removing acetaminophen, as compared with SO and, especially, CV. CS was also more efficient than SO and CV in the removal of salicylic acid [35]. However, in the case of diclofenac, SO was most capable than CS and CV [36], which evidences the importance of strain assortment for optimizing microalgae removal of specific pharmaceuticals from water. Zebrafish embryo exposure to the effluents from microalgae treatment further confirmed differences between strains. Although mortality rate was not affected by exposition to effluents, effects on the percentage of abnormalities was strain dependent. The effects were more remarkable in the sense CV > SO > CS, that is, contrary to the efficiency in the removal of acetaminophen. On the other hand, the reduction of acetaminophen toxic effects by treatment with these strains was also evaluated in this study. For this purpose, effluents from microalgae treatment were diluted at 1:1 with freshwater so to prevent the culture medium masking effects on embryo. Hence, comparing results in Figure 3 and Table 2 with those observed for 12,500 μg L^{-1} acetaminophen (Figure 1 and Table 1), it was clear that microalgae treatment by CS, CV and SO provided a reduction of effects on zebrafish embryo. This reduction was not evident for the shorter observation time-points (8 and 32 hpf) but was patent at 80 and 144 hpf. At the latter time-points, total abnormalities caused by exposure to the experimental solution of acetaminophen (12,500 μg L^{-1}) were 55% and 97%, respectively. Also, as can be observed in Table 1, the acetaminophen concentration of 12,500 μg L^{-1} caused excess of pigmentation in the 97% of the larvae at 144 hpf. Moreover, 65% remained in lateral position, 26% also showed spasms or involuntary movements and larvae were significantly larger than the control. In comparison, exposure to effluents from CS, which was the most capable in removing acetaminophen (67%, as shown in Figure 2), caused 17% and 37% of total abnormalities at 80 and 144 hpf, respectively, these being just restricted to lack of pigmentation. Meanwhile, for effluents from CV treatment, which was the least efficient, the average total abnormalities were 38% (80 hpf) and 63% (144 hpf), abnormalities including lack of pigmentation (32 hpf), excess of pigmentation (80 and 144 hpf), lateral position (144 hpf) and larval length (144 hpf). It is important to highlight that toxic effects on the embryo caused by exposure to microalgae-treated effluents were equivalent to those observed by acetaminophen experimental solutions with equivalent concentration. For instance, under exposure to the effluent from CV treatment (10,326 μg L^{-1} acetaminophen, at 1:1 dilution) the total abnormalities at 144 hpf were 63% (Figure 3), which was slightly lower than the abnormalities observed for the acetaminophen solution of 6250 μg L^{-1} (67% at 144 hpf, Figure 1). Among the different mechanisms that may be under the removal of pharmaceuticals by microalgae, biodegradation has been pointed as the most relevant [6]. Contrarily to physical treatments such as adsorption-based treatments, which also make possible the recuperation of the removed pharmaceuticals [37], those treatments involving their degradation result in the generation of TPs. Indeed, the possibility of these TPs being even more toxic than the original compounds raises some controversy around the application of degradation treatments for the removal of pharmaceuticals from water. The generation of TPs from microalgae removal of acetaminophen was not assessed in this work. Nonetheless, according to the obtained results, the reduction of acetaminophen concentration by microalgae occurred together with a proportional decrease of toxic effects on zebrafish embryo exposed to effluents. Therefore, if TPs were produced during treatment, their toxicity and/or concentration did not mean an increased toxicity for this model. Still, it remains unknown if the reduction of toxicity occurred in parallel with the decrease of acetaminophen concentration since these results refer just to the end of microalgae batch cultivation. Indeed, as highlighted by Vo et al. [9], different intermediates and/or end products may be formed depending on the degradation mechanism and pathway. In this sense, Zhou et al. [38] found that the maximum accumulated concentration of 1,4-benzoquinone, which is the main TP formed during the microbial transformation of acetaminophen in natural waters, occurred at different reaction times among the different water samples.

In the literature, there is a lack of works on the comprehensive evaluation of microalgae removal of pharmaceuticals, namely by coupling toxicity and analytic assessments. To our best knowledge, except for our previous study regarding the removal of diclofenac [39] there are no published works using toxicity tests on fish for this purpose. Compared with diclofenac removal by microalgae (67% to 99%) [39], the removal of acetaminophen was less efficient (17% to 67%). Still, in this work, the efficiency of CS in the removal of acetaminophen (67%) was the same than that previously found for diclofenac [39]. The lower removal of acetaminophen by microalgae as compared with other pharmaceuticals, including diclofenac, was also observed by Villar-Navarro et al. [40]. These authors [40] observed that the efficiency of a high rate algal pond (HRAP) removing pharmaceuticals was comparable to that of an activated-sludge based conventional process, except for acetaminophen and ibuprofen, which were less efficiently removed in the HRAP. However, the higher efficiency removing nutrients was highlighted as an advantage of HRAPs for their use as an alternative (or addition as tertiary treatment) to more conventional approaches based on activated sludge [40].

The necessity of complementing analytical information on the percent removal of pharmaceuticals in order to evaluate treatment efficiency has already been pointed out by some authors. Several approaches taken in this sense for the specific case of acetaminophen are shown in Table 3. Measuring oxygen uptake by bacteria in activated sludge was the strategy used by Ali et al. [41] to complement information based on chemical analysis on the removal of pharmaceuticals (including acetaminophen) by biosorption onto modified dead biomass of SO. These authors observed a drastic decrease of dissolved oxygen by 91% and 95% for wastewaters containing a pharmaceutical mixture of 125 mg/L and 250 mg/L, respectively. However, after the biosorption treatment, such wastewaters did not show significant differences from the control regarding the dissolved oxygen, which further proved the effectiveness of the treatment.

Although not for a microalgae-based treatment, but for the visible-light-driven photocatalytic removal of acetaminophen, Czech et al. [42] followed an analogous approach to that in this work and used *Vibrio fischeri* to assess the efficiency of the treatment for reducing water toxicity. These authors [42] concluded that photocatalytically treated model water containing acetaminophen revealed no toxicity to *Vibrio fischeri*. Similarly, Le et al. [43] coupled ecotoxicity (*Vibrio fischeri* 81.9%, Microtox®screening tests) and chemical analysis monitoring during the electro-fenton oxidation of acetaminophen thus establishing a very useful relationship between the degradation pathway of acetaminophen and the global toxicity evolution of the solution.

The application of microalgae cultures for the uptake of pharmaceuticals from water has been studied by several authors, mainly in the last decade. Coincidently with results in this work, Xiong et al. [44] found that different microalgae strains differed in their efficiency to remove a target pharmaceutical (enrofloxacin). Furthermore, these authors observed that the consortium of these strains displayed a comparable removal capacity to that of the most effective species. Even when microalgae are present as consortia in the aquatic environment, available data on the arrangement of efficient microalgae consortia to remove several distinct pharmaceuticals from different therapeutic classes are rather limited.

Table 3. Published approaches to evaluate treatment efficiency in the removal of acetaminophen by complementing analytic information by toxicity assessments.

Pharmaceutical/s	Initial Concentration ($\mu g\ L^{-1}$)	Treatment	Maximum Percent Removal	Toxicity Assessment	Reference
Mixture (tramadol, cefadroxil, acetaminophen, ciprofloxacin and ibuprofen)	125×10^3–250×10^3 (global concentration of the mixture)	Biosorption onto modified dead microalgae biomass	-	Oxygen uptake by bacteria in activated sludge	[41]
Acetaminophen	10×10^3	Visible-light-driven photocatalysis	82%	*Vibrio fischeri*	[42]
Acetaminophen	151×10^3	Electro-fenton oxidation	87% (mineralization)	*Vibrio fischeri*	[43]
Acetaminophen	25×10^3	Microalgae removal	67%	Zebrafish embryo	this work

Hydroxylation, side chain breakdown and ring cleavage have been reported as pathways for the biodegradation of aromatic pollutants by microalgae [45]. Nevertheless, more in-depth analyses need to be performed in order to explain differences between microalgae strains in the removal of specific pharmaceuticals. Furthermore, in view of the optimization of treatment duration, further studies should be done on the analysis of TPs and toxic effects throughout microalgae culture considering a broad spectrum of pharmaceuticals and strains. Finally, carrying out research on microalgae efficiency in the removal of pharmaceuticals using real wastewater matrices is a challenge to overcome in the near future.

5. Conclusions

In the considered range of concentrations (25 to 25,000 $\mu g\ L^{-1}$), experimental solutions of acetaminophen did not cause effects on the mortality of zebrafish embryos but significantly increased the total abnormalities at acetaminophen concentrations $\geq$2500 $\mu g\ L^{-1}$. Microalgae-based treatments by CS, CV and SO presented different efficiencies in the uptake of acetaminophen from water (17% to 67% at the end of the batch cultivation). Chemical analyses were coupled to zebrafish embryo bioassays for assessing the efficiency of these strains to remove the target pharmaceutical. It was evidenced that CS was the most efficient strain in reducing both acetaminophen concentration and toxic effects on zebrafish embryo. Furthermore, results confirmed that the effects of effluents from microalgae treatment were alike to those determined for experimental solutions with equivalent acetaminophen concentrations. It can be therefore concluded that microalgae removal of acetaminophen occurred together with a reduction of toxicity to zebrafish embryo, further supporting the potential application of microalgae for the removal of pharmaceuticals from water.

Supplementary Materials: The following are available online at http://www.mdpi.com/2073-4441/11/9/1929/s1, Figure S1: Periods of the embryo development of *Danio rerio*: (a,b) gastrula period; (c,d) pharyngula period; (e,f) larval stage. Note: Sketches have been taken from Kimmel et al. [28] and pictures from microscope. Table S1: Effects on zebrafish embryo exposed to effluents from microalgae treatments at a 1:3 dilution with freshwater. Note: Mean results (n = 12 for control; n = 6 for exposed groups) are shown together with SE. Results significantly different from control ($p \leq 0.05$) are in bold.

Author Contributions: Conceptualization, M.O. and M.M.S.; experimental design, M.M.S., C.E. and M.O.; methodology, M.M.S.; experimentation, C.E.; experimental support, T.T., T.N. and M.M.S.; analytic support,

R.N.C. and M.O.; resources, M.M.S.; data analysis and discussion, C.E., M.O. and M.M.S..; writing—original draft preparation, C.E., R.N.C. and M.O.; writing—review and editing, M.M.S., T.N. and M.O.

Funding: This research received funding by Fundação para a Ciência e a Tecnologia (FCT, Lisboa, Portugal), grant number IF/00314/2015. C.E. was supported by the Ministerio de Educación, Cultura y Deportes (Madrid, Spain), which awarded her with a grant for a short stay (EST15/00405) to be carried out at the Interdisciplinary Centre of Marine and Environmental Research (CIIMAR). T.N. was supported by the Fundação para a Ciência e a Tecnologia (FCT, Lisboa, Portugal), fellowship number SFRH/BPD/77912/2011. M.M.S. acknowledges funding from the project NOR-WATER "Poluentes emergentes nas águas da Galiza-Norte de Portugal: novas ferramentas para gestão de risco" (Reference: 0725_NOR_WATER_1_P), financed by "Programa de Cooperação Interreg Portugal/Espanha" (POCTEP) 2014–2020.

Acknowledgments: Thanks are due for financial support to Centre for Environmental and Marine Studies (CESAM) (UID/AMB/50017/2019), to FCT/MEC through national funds, and the co-funding by the European Regional Development Fund (FEDER), within the PT2020 Partnership Agreement and Compete 2020.

Conflicts of Interest: The authors declare no conflict of interest and that funders had no role in the design of the study; in the collection, analyses, or interpretation of data; in the writing of the manuscript, and in the decision to publish the results.

References

1. Quesada, H.B.; Baptista, A.T.A.; Cusioli, L.F.; Seibert, D.; Bezerra, C.D.O.; Bergamasco, R. Surface water pollution by pharmaceuticals and an alternative of removal by low-cost adsorbents: A review. *Chemosphere* **2019**, *222*, 766–780. [CrossRef] [PubMed]
2. Larsen, T.A.; Lienert, J.; Joss, A.; Siegrist, H. How to avoid pharmaceuticals in the aquatic environment. *J. Biotechnol.* **2004**, *113*, 295–304. [CrossRef] [PubMed]
3. Escapa, C.; Coimbra, R.N.; Nuevo, C.; Vega, S.; Paniagua, S.; García, A.I.; Calvo, L.F.; Otero, M. Valorization of Microalgae Biomass by Its Use for the Removal of Paracetamol from Contaminated Water. *Water* **2017**, *9*, 312. [CrossRef]
4. Kim, J.-Y.; Kim, H.-W. Photoautotrophic Microalgae Screening for Tertiary Treatment of Livestock Wastewater and Bioresource Recovery. *Water* **2017**, *9*, 192. [CrossRef]
5. Coimbra, R.N.; Escapa, C.; Vázquez, N.C.; Noriega-Hevia, G.; Otero, M. Utilization of Non-Living Microalgae Biomass from Two Different Strains for the Adsorptive Removal of Diclofenac from Water. *Water* **2018**, *10*, 1401. [CrossRef]
6. Xiong, J.-Q.; Kurade, M.B.; Jeon, B.-H. Can Microalgae Remove Pharmaceutical Contaminants from Water? *Trends Biotechnol.* **2018**, *36*, 30–44. [CrossRef] [PubMed]
7. Coimbra, R.N.; Escapa, C.; Otero, M. Comparative Thermogravimetric Assessment on the Combustion of Coal, Microalgae Biomass and Their Blend. *Energies* **2019**, *12*, 2962. [CrossRef]
8. Acién, F.G.; Molina, E.; Reis, A.; Torzillo, G.; Zittelli, G.C.; Sepúlveda, C.; Masojídek, J. Photobioreactors for the production of microalgae. In *Microalgae-based Biofuels and Bioproducts: From Feedstock Cultivation to End-Products*; González-Fernández, C., Muñoz, R., Eds.; Woodhead Publishing Series in Energy: Cambridge, UK, 2018; pp. 1–44.
9. Vo, H.N.P.; Ngo, H.H.; Guo, W.; Nguyen, T.M.H.; Liu, Y.; Liu, Y.; Nguyen, D.D.; Chang, S.W. A critical review on designs and applications of microalgae-based photobioreactors for pollutants treatment. *Sci. Total. Environ.* **2019**, *651*, 1549–1568. [CrossRef]
10. Judd, S.; Broeke, L.J.V.D.; Shurair, M.; Kuti, Y.; Znad, H. Algal remediation of CO_2 and nutrient discharges: A review. *Water Res.* **2015**, *87*, 356–366. [CrossRef]
11. Cuellar-Bermudez, S.P.; Aleman-Nava, G.S.; Chandra, R.; Garcia-Perez, J.S.; Contreras-Angulo, J.R.; Markou, G.; Muylaert, K.; Rittmann, B.E.; Parra-Saldivar, R. Nutrients utilization and contaminants removal. A review of two approaches of algae and cyanobacteria in wastewater. *Algal Res.* **2017**, *24*, 438–449. [CrossRef]
12. Tolboom, S.N.; Carrillo-Nieves, D.; Rostro-Alanis, M.D.J.; Quiroz, R.D.L.C.; Barceló, D.; Iqbal, H.M.; Parra-Saldivar, R. Algal-based removal strategies for hazardous contaminants from the environment–A review. *Sci. Total. Environ.* **2019**, *665*, 358–366. [CrossRef] [PubMed]
13. Wilkinson, J.; Hooda, P.S.; Barker, J.; Barton, S.; Swinden, J. Occurrence, fate and transformation of emerging contaminants in water: An overarching review of the field. *Environ. Pollut.* **2017**, *231*, 954–970. [CrossRef] [PubMed]

14. Bergheim, M.; Gminski, R.; Spangenberg, B.; Debiak, M.; Bürkle, A.; Mersch-Sundermann, V.; Kümmerer, K.; Gieré, R. Recalcitrant pharmaceuticals in the aquatic environment: a comparative screening study of their occurrence, formation of phototransformation products and their in vitro toxicity. *Environ. Chem.* **2014**, *11*, 431–444. [CrossRef]
15. Lai, K.M.; Scrimshaw, M.D.; Lester, J.N. Biotransformation and Bioconcentration of Steroid Estrogens by Chlorella vulgaris. *Appl. Environ. Microbiol.* **2002**, *68*, 859–864. [CrossRef]
16. Peng, F.-Q.; Ying, G.-G.; Yang, B.; Liu, S.; Lai, H.-J.; Liu, Y.-S.; Chen, Z.-F.; Zhou, G.-J. Biotransformation of progesterone and norgestrel by two freshwater microalgae (Scenedesmus obliquus and Chlorella pyrenoidosa): Transformation kinetics and products identification. *Chemosphere* **2014**, *95*, 581–588. [CrossRef]
17. Xiong, J.-Q.; Kim, S.-J.; Kurade, M.B.; Govindwar, S.; Abou-Shanab, R.A.; Kim, J.-R.; Roh, H.-S.; Khan, M.A.; Jeon, B.-H. Combined effects of sulfamethazine and sulfamethoxazole on a freshwater microalga, Scenedesmus obliquus: Toxicity, biodegradation, and metabolic fate. *J. Hazard. Mater.* **2019**, *370*, 138–146. [CrossRef] [PubMed]
18. Stadler, L.B.; Ernstoff, A.S.; Aga, D.S.; Love, N.G. Micropollutant Fate in Wastewater Treatment: Redefining "Removal". *Environ. Sci. Technol.* **2012**, *46*, 10485–10486. [CrossRef]
19. De Laurentiis, E.; Prasse, C.; Ternes, T.A.; Minella, M.; Maurino, V.; Minero, C.; Sarakha, M.; Brigante, M.; Vione, D. Assessing the photochemical transformation pathways of acetaminophen relevant to surface waters: Transformation kinetics, intermediates, and modelling. *Water Res.* **2014**, *53*, 235–248. [CrossRef]
20. Caballero, M.V.; Candiracci, M. Zebrafish as screening model for detecting toxicity and drugs efficacy. *J. Unexplored Med. Data* **2018**, *3*, 1–14. [CrossRef]
21. Carlsson, G.; Patring, J.; Kreuger, J.; Norrgren, L.; Oskarsson, A. Toxicity of 15 veterinary pharmaceuticals in zebrafish (*Danio rerio*) embryos. *Aquat. Toxicol.* **2013**, *126*, 30–41. [CrossRef]
22. Dai, Y.J.; Jia, Y.F.; Chen, N.; Bian, W.P.; Li, Q.K.; Ma, Y.B.; Chen, Y.L.; Pei, D.S. Zebrafish as a model system to study toxicology. *Environ. Toxicol. Chem.* **2014**, *33*, 11–17. [CrossRef] [PubMed]
23. Torres, T.; Cunha, I.; Martins, R.; Santos, M.M. Screening the Toxicity of Selected Personal Care Products Using Embryo Bioassays: 4-MBC, Propylparaben and Triclocarban. *Int. J. Mol. Sci.* **2016**, *17*, 1762. [CrossRef] [PubMed]
24. Soares, J.; Coimbra, A.; Reis-Henriques, M.; Monteiro, N.; Vieira, M.; Oliveira, J.; Guedes-Dias, P.; Fontaínhas-Fernandes, A.; Parra, S.S.; Carvalho, A.P.; et al. Disruption of zebrafish (*Danio rerio*) embryonic development after full life-cycle parental exposure to low levels of ethinylestradiol. *Aquat. Toxicol.* **2009**, *95*, 330–338. [CrossRef] [PubMed]
25. Ribeiro, S.; Torres, T.; Martins, R.; Santos, M.M. Toxicity screening of Diclofenac, Propranolol, Sertraline and Simvastatin using *Danio rerio* and *Paracentrotus lividus* embryo bioassays. *Ecotoxicol. Environ. Saf.* **2015**, *114*, 67–74. [CrossRef] [PubMed]
26. Mann, J.E.; Myers, J. On pigments, growth, and photosynthesis of phaeodactylum tricornutum. *J. Phycol.* **1968**, *4*, 349–355. [CrossRef] [PubMed]
27. OECD. *OECD Guidelines for the Testing of Chemicals*; Test No. 236: Fish Embryo Acute Toxicity (FET) Test; OECD Publishing: Paris, France, 2013. Available online: https://read.oecd-ilibrary.org/environment/test-no-236-fish-embryo-acute-toxicity-fet-test_9789264203709-en#page1 (accessed on 12 September 2019).
28. Kimmel, C.B.; Ballard, W.W.; Kimmel, S.R.; Ullmann, B.; Schilling, T.F. Stages of embryonic development of the zebrafish. *Dev. Dyn.* **1995**, *203*, 253–310. [CrossRef] [PubMed]
29. Escapa, C.; Coimbra, R.; Paniagua, S.; García, A.; Otero, M.; Coimbra, R. Nutrients and pharmaceuticals removal from wastewater by culture and harvesting of Chlorella sorokiniana. *Bioresour. Technol.* **2015**, *185*, 276–284. [CrossRef]
30. Nogueira, A.F.; Pinto, G.; Correia, B.; Nunes, B. Embryonic development, locomotor behavior, biochemical, and epigenetic effects of the pharmaceutical drugs paracetamol and ciprofloxacin in larvae and embryos of *Danio rerio* when exposed to environmental realistic levels of both drugs. *Environ. Toxicol.* **2019**, in press. [CrossRef]
31. Xia, L.; Zheng, L.; Zhou, J.L. Effects of ibuprofen, diclofenac and paracetamol on hatch and motor behavior in developing zebrafish (*Danio rerio*). *Chemosphere* **2017**, *182*, 416–425. [CrossRef]
32. Pandya, M.; Patel, D.; Rana, J.; Patel, M.; Khan, N. Hepatotoxicity by Acetaminophen and Amiodarone in Zebrafish Embryos. *J. Young Pharm.* **2016**, *8*, 50–52. [CrossRef]

33. Peng, H.-C.; Wang, Y.-H.; Wen, C.-C.; Wang, W.-H.; Cheng, C.-C.; Chen, Y.-H. Nephrotoxicity assessments of acetaminophen during zebrafish embryogenesis. *Comp. Biochem. Physiol. Part C: Toxicol. Pharmacol.* **2010**, *151*, 480–486. [CrossRef] [PubMed]
34. David, A.; Pancharatna, K. Effects of acetaminophen (paracetamol) in the embryonic development of zebrafish, *Danio rerio*. *J. Appl. Toxicol.* **2009**, *29*, 597–602. [CrossRef] [PubMed]
35. Escapa, C.; Coimbra, R.; Paniagua, S.; García, A.; Otero, M. Paracetamol and salicylic acid removal from contaminated water by microalgae. *J. Environ. Manag.* **2017**, *203*, 799–806. [CrossRef] [PubMed]
36. Escapa, C.; Coimbra, R.; Paniagua, S.; García, A.; Otero, M. Comparative assessment of diclofenac removal from water by different microalgae strains. *Algal Res.* **2016**, *18*, 127–134. [CrossRef]
37. Ramos, A.M.; Otero, M.; Rodrigues, A.E. Recovery of Vitamin B12 and cephalosporin-C from aqueous solutions by adsorption on non-ionic polymeric adsorbents. *Sep. Purif. Technol.* **2004**, *38*, 85–98. [CrossRef]
38. Zhou, C.; Zhou, Q.; Zhang, X. Transformation of acetaminophen in natural surface water and the change of aquatic microbes. *Water Res.* **2019**, *148*, 133–141. [CrossRef] [PubMed]
39. Escapa, C.; Torres, T.; Neuparth, T.; Coimbra, R.N.; García, A.I.; Santos, M.M.; Otero, M. Zebrafish embryo bioassays for a comprehensive evaluation of microalgae efficiency in the removal of diclofenac from water. *Sci. Total. Environ.* **2018**, *640*, 1024–1033. [CrossRef]
40. Villar-Navarro, E.; Baena-Nogueras, R.M.; Paniw, M.; Perales, J.A.; Lara-Martin, P.A. Removal of pharmaceuticals in urban wastewater: High rate algae pond (HRAP) based technologies as an alternative to activated sludge based processes. *Water Res.* **2018**, *139*, 19–29. [CrossRef]
41. Ali, M.E.; El-Aty, A.M.A.; Badawy, M.I.; Ali, R.K. Removal of pharmaceutical pollutants from synthetic wastewater using chemically modified biomass of green alga Scenedesmus obliquus. *Ecotoxicol. Environ. Saf.* **2018**, *151*, 144–152. [CrossRef]
42. Czech, B.; Jośko, I.; Oleszczuk, P. Ecotoxicological evaluation of selected pharmaceuticals to Vibrio fischeri and Daphnia magna before and after photooxidation process. *Ecotoxicol. Environ. Saf.* **2014**, *104*, 247–253. [CrossRef]
43. Le, T.X.H.; Van Nguyen, T.; Yacouba, Z.A.; Zoungrana, L.; Avril, F.; Nguyen, D.L.; Petit, E.; Mendret, J.; Bonniol, V.; Bechelany, M.; et al. Correlation between degradation pathway and toxicity of acetaminophen and its by-products by using the electro-Fenton process in aqueous media. *Chemosphere* **2017**, *172*, 1–9. [CrossRef] [PubMed]
44. Xiong, J.-Q.; Kurade, M.B.; Jeon, B.-H. Ecotoxicological effects of enrofloxacin and its removal by monoculture of microalgal species and their consortium. *Environ. Pollut.* **2017**, *226*, 486–493. [CrossRef] [PubMed]
45. Semple, K.T.; Cain, R.B.; Schmidt, S. Biodegradation of aromatic compounds by microalgae. *FEMS Microbiol. Lett.* **1999**, *170*, 291–300. [CrossRef]

MDPI
St. Alban-Anlage 66
4052 Basel
Switzerland
Tel. +41 61 683 77 34
Fax +41 61 302 89 18
www.mdpi.com

Water Editorial Office
E-mail: water@mdpi.com
www.mdpi.com/journal/water